LaTeX Beginner's Guide

Third Edition

Write research papers, theses, and presentations with professional formatting, math, and citations

Stefan Kottwitz

<packt>

LaTeX Beginner's Guide

Third Edition

Portfolio Director: Pavan Ramchandani

Relationship Lead: Alok Dhuri

Content Engineer: Rounak Kulkarni

Technical Editor: Vidhisha Patidar

Copy Editor: Safis Editing

Indexer: Rekha Nair

Proofreader: Rounak Kulkarni

Production Designer: Salma Patel

Growth Lead: Nivedita Singh

First published: March 2011

Second edition: August 2021

Third edition: February 2026

Production reference: 01230226

Published by Packt Publishing Ltd.

Grosvenor House

11 St Paul's Square

Birmingham

B3 1RB, UK.

ISBN 978-1-80580-457-4

`www.packtpub.com`

To the members of TUG and DANTE for supporting TeX and LaTeX development, infrastructure, and education. To all the helpers on internet forums for their tireless support for LaTeX beginners.

– Stefan Kottwitz

Contributors

About the author

Stefan Kottwitz studied mathematics in Jena and Hamburg. He works as a network and IT security engineer for Lufthansa Industry Solutions.

For many years, he has been providing LaTeX support on online forums. He maintains the web forums `LaTeX.org` and `goLaTeX.de` and the Q&A sites `TeXwelt.de` and `TeXnique.fr`. He runs the TeX graphics gallery sites `TeXample.net`, `TikZ.net`, and `PGFplots.net`, the `TeXlive.net` online compiler, the `TeXdoc.org` service, and the `CTAN.net` software mirror. He is a moderator on TeX Stack Exchange, part of the Stack Overflow network, and on `matheplanet.com`. He publishes ideas and news from the TeX world on his blogs `LaTeX.net` and `TeX.co`.

Besides *LaTeX Beginner's Guide*, he wrote *LaTeX Cookbook* and *LaTeX Graphics with TikZ*, both also translated and published in Japanese.

About the reviewers

Izaak Neutelings earned his master's and Ph.D. degrees from the **University of Zurich (UZH)**. He is currently engaged in fundamental research in experimental particle physics at the CMS experiment in CERN, where he focuses on searching for physics beyond the Standard Model in proton-proton collisions, with a specialization in tau-lepton final states.

He previously served as a technical reviewer for *LaTeX Graphics with TikZ* and *LaTeX Cookbook*. His extensive use of LaTeX includes writing lecture notes for introductory physics courses at UZH, fully illustrated with TikZ figures. He is a primary contributor to the websites `TikZ.net` and `FeynM.net`, which both feature large galleries of graphics that visitors reproduce with the provided LaTeX code.

Joseph Wright is the author of several widely used LaTeX packages and is a member of the LaTeX Project team. Joseph is a chemist by training, and in his day job is a university lecturer in inorganic chemistry.

Table of Contents

Chapter 2: Formatting Text and Creating Macros 27

Preface

LaTeX is high-quality, open source typesetting software that produces professional-quality print and PDF output. However, as LaTeX is a powerful and complex tool, getting started can be intimidating, and specific aspects such as layout modifications can seem rather complicated. Using Microsoft Word or other word-processing software may seem more straightforward, but once you've become acquainted with it, LaTeX's capabilities far outweigh any initial difficulties. This book guides you through these challenges and makes it easy to begin with LaTeX. If you are writing mathematical, scientific, or technical papers, this is the perfect book for you.

LaTeX Beginner's Guide, Third Edition, offers you a practical introduction to LaTeX, guiding you through the essential steps of LaTeX, from installing LaTeX to formatting, justification, and page design, including presentation slides. You will learn how to use macros and styles to maintain a consistent document structure while reducing typing. Beginning with installation and basic usage, you will learn how to typeset documents that include professional-looking tables, figures, mathematical formulas, and common book elements such as bibliographies, glossaries, and indexes. Lots of step-by-step examples start with fine-tuning text, formulas, and page layout, and proceed with managing complex documents and using modern PDF features. Detailed information about online resources such as software archives, web forums, and online compilers complements this introductory guide.

Finally, we'll look under the hood at the TeX engines and use artificial intelligence tools to support our work.

Who this book is for

If you are about to write mathematical or scientific papers, seminar handouts, or even plan to write a thesis, then this book offers you a fast-paced and practical introduction. Those studying at school and university as mathematicians or physicists will benefit greatly, as will engineers and humanities students. Anybody with high expectations who plans to write a paper or a book will be delighted by this high-quality, stable software.

What this book covers

Chapter 1, Getting Started with LaTeX, introduces LaTeX and explains its benefits. It guides you through the download and installation of a comprehensive LaTeX distribution and shows you how to create your first LaTeX document. It also introduces Overleaf, an online LaTeX editor. You'll also learn how to access package documentation.

Chapter 2, Formatting Text and Creating Macros, explains how to vary fonts, shapes, and text styles. It covers centering and justification of paragraphs, and how we can improve line breaks and hyphenation. It introduces logical formatting and explains how to define macros and use environments and packages.

Chapter 3, Designing Pages, shows how to adjust margins and line spacing, and how to work with portrait, landscape, and two-column layouts. It shows how to create dynamic headers and footers, control page breaks, and use footnotes. Along the way, you'll also learn how to redefine existing commands and make effective use of class options.

Chapter 4, Creating Lists, covers bulleted, numbered, and definition lists. It shows how to choose bullet and numbering styles and how to fine-tune the overall layout of lists.

Chapter 5, Creating Tables, shows you how to create professional-looking tables and goes deep into formatting details.

Chapter 6, Including Images, explains how to include external images with captions and how LaTeX handles figure placement automatically, with options to adjust it when needed.

Chapter 7, Using Cross-References, introduces smart referencing for sections, footnotes, tables, figures, and numbered environments.

Chapter 8, Managing Contents, Indexes, and Bibliography, explains how to create and customize the table of contents and the lists of figures and tables. It also shows how to cite sources, build bibliographies, and create an index.

Chapter 9, Writing Math Formulas, introduces mathematical typesetting in depth. It starts with basic formulas, then moves on to displayed and numbered equations and the alignment of multi-line expressions. You'll learn how to typeset math symbols such as roots, arrows, Greek letters, and operators, and how to build more complex structures, such as fractions, stacked expressions, and matrices.

Chapter 10, Using Fonts, introduces font selection in LaTeX and shows Roman, sans-serif, and typewriter fonts in various shapes and styles.

Chapter 11, Developing Large Documents, shows how to manage larger projects by splitting them into multiple files. You'll learn how to build complex documents from smaller parts, handle front and back matter with separate page numbering, and create custom title pages. Working through an example book, you'll also get familiar with templates, preparing you to write your own thesis, book, or report.

Chapter 12, Using Hyperlinks and Designing Headings, explains how to enrich PDF documents with hyperlinks, bookmarks, and metadata. It also shows how to customize chapter and section headings.

Chapter 13, Creating Presentations, walks through building slides with LaTeX, from incremental reveals to column and block layouts, and generating handouts.

Chapter 14, Troubleshooting, focuses on solving problems. You'll learn about typical LaTeX errors and warnings and how to read and use LaTeX's messages to track down and correct issues.

Chapter 15, Using Online Resources, shows you where to find reliable LaTeX information on the internet. We'll look at forums and Q&A sites, explore software archives and TeX user group pages, and point you to mailing lists, Usenet groups, and graphics galleries. The chapter also highlights where to download LaTeX editors and where LaTeX users meet on blogs and X.com.

Chapter 16, Exploring Technology: From Engines to AI, takes a look under the hood at TeX engines, compares them, and discusses when to use which one. The chapter also covers AI tools that can improve your coding, writing, and formatting workflows and concludes with recommended reading.

To get the most out of this book

You need access to a computer with LaTeX on it. An online connection would be helpful regarding installation and updates. We can install LaTeX on most operating systems, so you can use Windows, Linux, macOS, or Unix.

This book uses the freely available TeX Live distribution, which runs on all the mentioned platforms. You just need an internet connection or the TeX Live DVD to install it. In the book, we use the cross-platform editor TeXworks, but you can use any editor you like.

Without installing LaTeX, you can work with the code examples at `https://latexguide.org`, which comes with an online compiler.

If you are using the digital version of this book, we advise you to type the code yourself or access the code from the book's GitHub repository (a link is available in the next section). Doing so will help you avoid potential errors when copying and pasting code.

Download the example code files

The code bundle for the book is hosted on GitHub at `https://github.com/PacktPublishing/LaTeX-Beginner-s-Guide-Third-Edition`. We also have other code bundles from our rich catalog of books and videos available at `https://github.com/PacktPublishing`. Check them out!

Download the color images

We also provide a PDF file that has color images of the screenshots/diagrams used in this book. You can download it here: `https://packt.link/gbp/9781805804574`.

Conventions used

A number of text conventions are used throughout this book.

`CodeInText`: Indicates code words in text, database table names, folder names, filenames, file extensions, pathnames, dummy URLs, user input, and Twitter handles. For example: "The first line starts with the `\documentclass` command."

A block of code is set as follows:

```
\begin{name}[optional argument]{argument}
    ...
\end{name}
```

When we wish to draw your attention to a particular part of a code block, the relevant lines or items are set in bold:

```
\documentclass{article}
\newcommand{\keyword}[2][\bfseries]{{#1#2}}
\begin{document}
```

Any command-line input or output is written as follows:

```
texdoc beamer
```

Screentext: Indicates a new term, an important word, or words that you see on the screen. For instance, words in menus or dialog boxes appear in the text like this. For example: "Click **Typeset** and check out the result."

Warnings or important notes appear like this.

Tips and tricks appear like this.

Get in touch

Feedback from our readers is always welcome.

General feedback: If you have questions about any aspect of this book or general feedback, please email us at `customercare@packt.com` and mention the book's title in the subject line of your message.

Errata: Although we have taken every care to ensure the accuracy of our content, mistakes do happen. If you have found a mistake in this book, we would be grateful if you could report it to us. Please visit `http://www.packt.com/submit-errata`, click **Submit Errata**, and fill in the form.

Piracy: If you come across any illegal copies of our works in any form on the internet, we would be grateful if you could provide the location address or website name. Please contact us at `copyright@packt.com` with a link to the material.

If you are interested in becoming an author, and you have expertise in a topic and would like to write or contribute to a book, please visit `http://authors.packt.com/`.

Share your thoughts

Once you've read *LaTeX Beginner's Guide, Third Edition*, we'd love to hear your thoughts! Scan the QR code below to go straight to the Amazon review page for this book and share your feedback.

`https://packt.link/r/180580457X`

Your review is important to us and the tech community and will help us make sure we're delivering excellent quality content.

Free benefits with your book

This book comes with free benefits to support your learning. Activate them now for instant access (see the *How to unlock* section for instructions).

Here's a quick overview of what you can instantly unlock with your purchase:

How to unlock

Scan the QR code (or go to `packtpub.com/unlock`). Search for this book by name, confirm the edition, and then follow the steps on the page.

UNLOCK NOW

Note: Keep your invoice handy. Purchases made directly from Packt don't require one

1

Getting Started with LaTeX

You are familiar with **word processing** software: you type some text, and what you see on the screen is what you get on the page. In contrast, LaTeX is a **typesetting** system: you write plain text with **markup commands**, and LaTeX turns it into polished output. It produces high-quality prints and PDF files based on sophisticated algorithms for justification, hyphenation, text alignment, whitespace balancing, and figure placement. It comes with predefined formatting styles for headings, margins, and general page layout, which you can customize.

Now that you're ready to move beyond those "what you see is what you get" word processors and start using a system designed for precision and consistency, you're in the right place. Let's take the first steps together.

This book will guide you along the way, helping you get the most out of LaTeX's capabilities. First, we'll look at its strengths and a few challenges before setting up the tools you'll need.

In this chapter, we'll get acquainted with LaTeX, what it is, and how to install and use it. Specifically, we'll cover the following:

- Getting the LaTeX idea
- Installing and using LaTeX
- Working with LaTeX online
- Accessing documentation

By the end of this chapter, you'll have LaTeX up and running, know how to edit and typeset a basic document, and understand where to find further documentation.

Let's get started.

Your purchase includes a free PDF copy + exclusive extras

Your purchase includes a DRM-free PDF copy of this book, 7-day trial to the Packt+ library (no credit card required), and additional exclusive extras. See the *Free benefits with your book* section in the *Preface* to unlock them instantly and maximize your learning.

Technical requirements

We'll focus on the Windows operating system here, but you can also install LaTeX on macOS, Linux, and other systems.

A full installation typically requires around 8 GB of disk space on every system.

If you prefer not to install anything, you can use an online platform such as Overleaf, which runs entirely in your browser and requires a constant internet connection. We'll cover Overleaf at the end of this chapter.

All code examples from this book are available for download on GitHub at `https://github.com/PacktPublishing/LaTeX-Beginner-s-Guide-Third-Edition`.

On the book's website, `https://latexguide.org`, you can read, edit, and compile all the code examples directly in your browser—no installation required. All you need is a device with a modern web browser and JavaScript enabled, whether it's a PC, laptop, tablet, or smartphone.

Getting the LaTeX idea

LaTeX is free and open source software for typesetting documents. It's a **document preparation system** based on **markup commands**, where you describe the structure and formatting of your content using plain text commands.

It was initially written by Leslie Lamport in the 1980s and is built on top of the **TeX** typesetting engine by Donald Knuth, who started developing it in the late 1970s. It has a long history; you can read about it at `https://tug.org/whatis.html`. As Knuth derived the name from "tau epsilon chi", abbreviating the Greek word "τέχνη" for art, craft, or technology, people commonly pronounce it like "Tech." Some prefer saying "tek" as in "technology." What it shouldn't sound like is "teks." For LaTeX, it's similar; we pronounce it like "Lahtech" or "Laytech."

Let's look at how we can make the most out of LaTeX.

Benefits of LaTeX

LaTeX is particularly well-suited for scientific and technical documents. LaTeX's superior typesetting of mathematical formulas is legendary. If you're a student or a scientist, LaTeX is by far the best choice. Even if you don't need its scientific capabilities, LaTeX has much to offer—it produces outstanding, high-quality output and is incredibly stable. It handles large and complex documents with ease.

Some of LaTeX's remarkable strengths are its cross-referencing capabilities, automatic numbering, and the generation of lists of contents, figures, and tables, indexes, glossaries, and bibliographies. It is multilingual with language-specific features and can use even advanced PDF features.

While it's the perfect choice for students' theses and advanced scientists' publishing, LaTeX is incredibly flexible—there are templates for letters, CVs, presentations, bills, philosophy books, law texts, music scores, and even chess game notation. Hundreds of LaTeX developers and power users have written thousands of templates, styles, and valuable tools for every possible purpose. These are collected and categorized online on archive servers.

You can take advantage of LaTeX's impressive high quality by starting with its default styles and relying on its built-in smart formatting. But you can customize nearly every aspect if you need something more tailored. The LaTeX community has developed a wide range of extensions covering almost every formatting requirement you might encounter.

Virtues of open source

LaTeX is entirely **open source**, with freely readable code for everyone. This enables everybody to study and adjust everything, from the core of LaTeX to the latest extension packages. But what does this mean for you as a beginner? A vast, active, and supportive LaTeX community has many friendly, helpful people. Even if you cannot benefit from the open source code directly, they can read the source and assist you. Just join a LaTeX web forum and ask your questions there. Helpers will, if necessary, dig into LaTeX sources and, in all probability, find a solution for you, sometimes by recommending a suitable package, often providing a redefinition of a default command.

Today, we benefit from more than 40 years of ongoing development by the LaTeX community. The open source philosophy made huge progress possible, as every user is invited to study, improve, and extend the software. *Chapter 15, Using Online Resources*, will connect you with the online community.

Separation of form and content

A core idea behind LaTeX is that the author shouldn't be distracted by formatting while writing. Instead of manually adjusting the appearance of text, you focus on the structure and meaning of your content. For example, you use a LaTeX command, marking it as a chapter heading rather than writing a chapter title in big, bold letters. You can let LaTeX apply the styling to the heading or define what the headings will look like in the document's settings—just once for the whole document. LaTeX relies on style files called **classes** and **packages**, making it easy to design and consistently modify the entire document's appearance and details.

Portability

LaTeX is available for nearly every major operating system, such as Windows, Linux, and macOS. It uses plain text files for documents and styles, which are readable and editable on all operating systems and versions. That means your documents are portable and LaTeX will produce the same output on each system. Several LaTeX software collections contain compiler engines, tools, and styles. We call them **TeX distributions**. We will focus on the **TeX Live** distribution since this one is available for Windows, Linux, and macOS. On the Mac, the customized TeX Live version is called **MacTeX**.

LaTeX doesn't have its own graphical user interface, which is one reason it's so portable. You can choose any text editor. However, there are many LaTeX-specific editors for every major operating system. Some, such as **TeXworks**, run on multiple systems, which is one of the reasons we will use it in our book. Another significant reason is that it's simple and best suited for beginners.

LaTeX generates **PDF** output, which is ideal for sharing since it's printable and readable on most computers and looks the same regardless of the operating system. Besides PDF, it supports **DVI**, **PostScript**, and **HTML** output, preparing the ground for distribution in print and online, such as on personal computers, electronic book readers such as the Kindle, and smartphones. To sum up, LaTeX is portable in three ways: your source files, its implementation, and the final output.

Protection for your work

LaTeX documents are stored in a human-readable text format, not in some proprietary word processing format exclusive to a specific software vendor, which may even be different between versions of the same software.

Try opening a 20-year-old document written with a commercial word processor. Even if it can open the file, its visual appearance would be altered, the formatting may be off, or the display may be wrong. LaTeX promises that the document will always be readable and consistently produce the same output. Despite ongoing development, it maintains strong backward compatibility.

Word processor documents can carry viruses or malicious macros that can harm your computer system and data. Viruses do not infect LaTeX documents because they can hardly hide in a plain text file.

How to get started with LaTeX

LaTeX can seem challenging at first, but this book will help you master it step by step, topic by topic.

Writing LaTeX looks like programming, but don't be intimidated. You'll quickly become familiar with the most commonly used commands, and text editors with auto-completion and syntax highlighting will support you. Some even offer you menus and dialog boxes to insert commands.

Worried it might take a long time to produce usable results? Don't be; this book will give you a quick start with plenty of hands-on examples, so you'll learn by doing and build confidence. Many more examples can be found on the internet. In *Chapter 15, Using Online Resources*, we will explore online resources. There are LaTeX help forums where you get answers to your questions. One of them has a section dedicated specifically to readers of this book and its companions: `https://latex.org/books`.

Approaches to working with LaTeX

There are two main ways to start working with LaTeX:

- The traditional approach is to install LaTeX on your own computer. It's fairly simple, and we will walk through installing it on Windows in the *Installing and using LaTeX* section.
- Another way is to use LaTeX entirely online in your browser. No installation is needed; you only need an internet-connected computer, tablet, or phone. We will explore this option in the *Working with LaTeX online using Overleaf* section at the end of this chapter.

We'll begin by setting up LaTeX locally on our computer. If you like, you could skip this moment and jump ahead to the *Working with LaTeX online using Overleaf* section, then decide which approach you would like to take.

Installing and using LaTeX

Let's start by installing the LaTeX distribution **TeX Live**. It's available for Windows, Linux, macOS (as **MacTeX**), and other Unix-like operating systems. TeX Live is actively developed and well-maintained, making it a reliable choice across platforms.

Another excellent and user-friendly TeX distribution for Windows is **MiKTeX**. It's easy to install, like a common Windows application. You can download it from `https://miktex.org`. Visit `https://latexguide.org/distributions` for a detailed, up-to-date comparison.

You can install TeX Live in one of two ways: in **single-user mode** only for your account, or system-wide for all users on a computer. The latter is called **admin mode**. It requires running the installation as an administrator: either log in with an administrator account or right-click on the **Install program** and choose **Run as administrator**.

If you are the only person using your computer, a single-user installation is recommended.

Let's start by visiting the TeX Live home page and reviewing the available installation options. To do this, open the TeX Live home page using `https://tug.org/texlive/`:

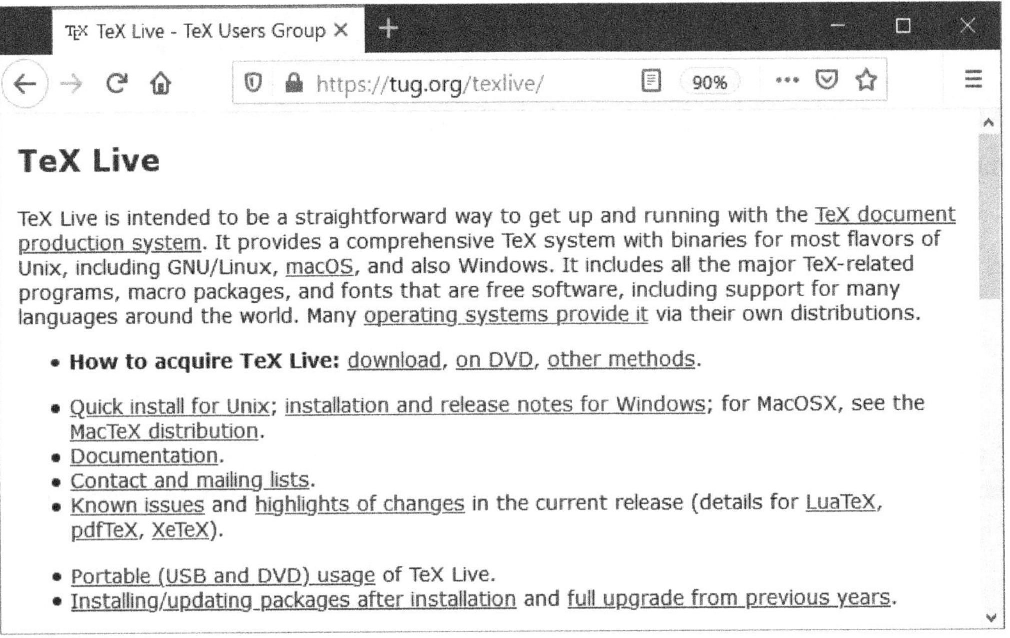

Figure 1.1 – TeX Live home page

Feel free to explore the home page in more detail to study the information offered there, though, in this book, we'll focus on two types of installation:

- Installing TeX Live **online** using the net installer wizard; this method requires a constant internet connection for typically about two hours or longer
- Installing TeX Live **offline**; this way starts with a considerable download, but then we can continue without having the internet available, with less risk of interruption

Before starting the installation, let's have a look at LaTeX files and packaging conventions with different granularity:

- A LaTeX **class** file defines the overall structure and layout of a document, such as sectioning behavior, default fonts, margins, and more. Class files have the `.cls` filename extension.
- A **package**, also called a **style** file, is a single LaTeX file with some macros to add specific features or provide a particular look and document style. Its filename extension is `.sty`.
- A **bundle** is a set of related files with a similar purpose, such as classes, packages, and support files.
- A **collection** is a more extensive set of packages for a field of interest. That can be, for example, an extensive set of math and natural science packages, music packages, or graphics-related packages.
- A **scheme** is a LaTeX installation of a specific size. It can be **minimal** (the smallest necessary to work), **basic** (commonly needed stuff), or **full** (everything available).

With this understanding, we can now install LaTeX. The easiest option is to install everything fully; that's the **full scheme**. This ensures that all packages are available from the start, so you won't encounter missing features later.

Let's check out two installation methods on a Windows PC. The first installs TeX Live over the internet, requiring a good internet connection. If you don't have a good internet connection, go to the next section, *Installing TeX Live offline*.

Installing TeX Live using the net installer wizard

We will download the TeX Live net installer and install the complete TeX Live distribution on our computer. To do this, follow these steps:

1. Click on **download**, as seen in *Figure 1.1*, or go to `https://tug.org/texlive/acquire-netinstall.html`:

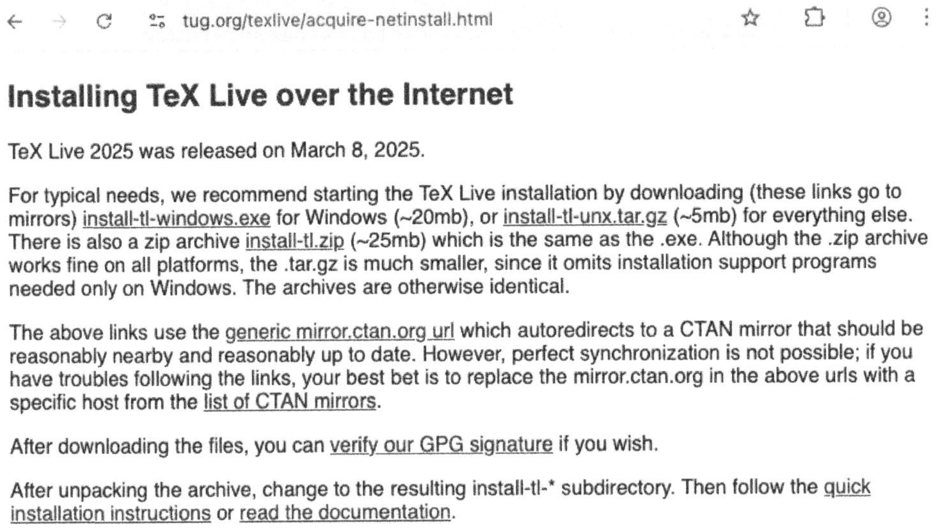

tug.org/texlive/acquire-netinstall.html

Installing TeX Live over the Internet

TeX Live 2025 was released on March 8, 2025.

For typical needs, we recommend starting the TeX Live installation by downloading (these links go to mirrors) install-tl-windows.exe for Windows (~20mb), or install-tl-unx.tar.gz (~5mb) for everything else. There is also a zip archive install-tl.zip (~25mb) which is the same as the .exe. Although the .zip archive works fine on all platforms, the .tar.gz is much smaller, since it omits installation support programs needed only on Windows. The archives are otherwise identical.

The above links use the generic mirror.ctan.org url which autoredirects to a CTAN mirror that should be reasonably nearby and reasonably up to date. However, perfect synchronization is not possible; if you have troubles following the links, your best bet is to replace the mirror.ctan.org in the above urls with a specific host from the list of CTAN mirrors.

After downloading the files, you can verify our GPG signature if you wish.

After unpacking the archive, change to the resulting install-tl-* subdirectory. Then follow the quick installation instructions or read the documentation.

Figure 1.2 – Installation instructions

2. Download the executable installer program, `install-tl-windows.exe`, and run the program.
3. Confirm the installation mode (**Single-user** or **Administrator**), click **Next**, and then **Install**.
4. The net installer will automatically detect your operating system language. You can change the language of the **graphical user interface** (**GUI**) by clicking on **GUI language** in the menu of the window that opens, as we can see in *Figure 1.3*:

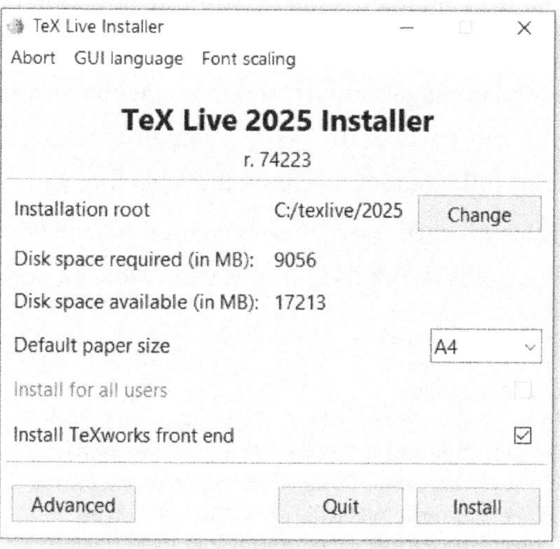

Figure 1.3 – TeX Live Installer

5. You can change the installation root: the location of all TeX Live installed files on your hard drive. That default installation usually works well, though you can click on **Advanced** to adjust specific options, as you can see here in *Figure 1.4*:

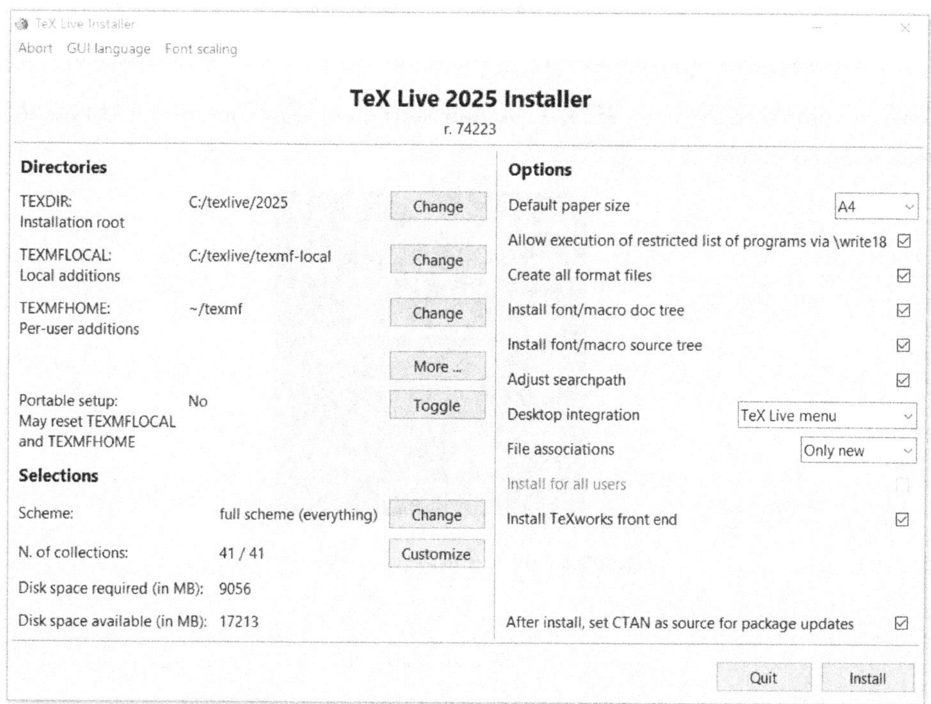

Figure 1.4 – TeX Live Installer advanced options

6. You can change the **Scheme** setting (options include **full**, **medium**, and **small**) and customize the number of software collections, such as installed formats, fonts, styles, graphics packages, editor, language support, and more. As the recommended options are already the most significant parts of the installation, unchecking some collections won't save much space. The full scheme is recommended so that you won't miss anything later.
7. Click on **Install** to proceed. Now, there's no input needed for a long time, and you can sit back while thousands of TeX packages are downloaded and installed:

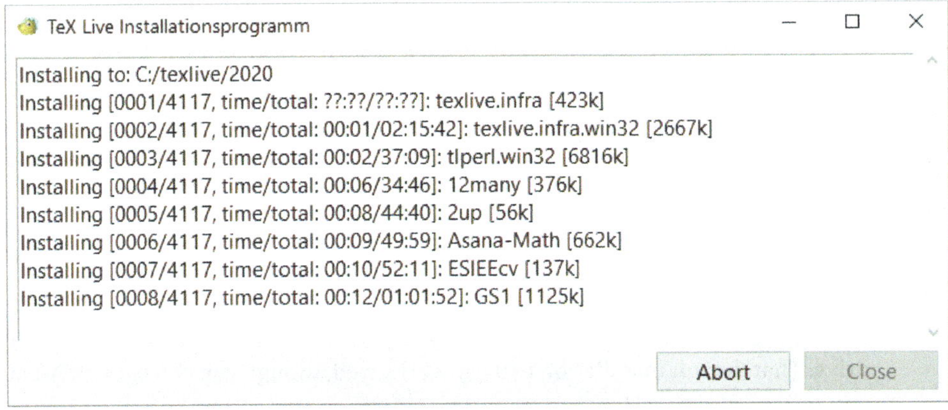

Figure 1.5 – Installation progress

8. Finally, you will get a welcome message. Click **Close** to finish the installation.

You have completed the setup of TeX Live. Your **Start** menu now contains a **TeX Live 2025** folder containing six programs:

Figure 1.6 – TeX Live in the Windows Start menu

Let's briefly look at each of the programs:

- **DVIOUT DVI viewer**: This is a viewer program for the classic LaTeX output format DVI (today, most people choose PDF output, so you probably won't need this).
- **TeX Live command-line**: Use this if you would like to run other TeX Live programs at the command line.
- **TeX Live documentation**: This opens the TeX Live manual in your web browser.
- **TeX Live Manager**: This is your tool for package management (for example, for installing and updating LaTeX packages).
- **TeXworks editor**: This is an editor that was developed to create LaTeX documents comfortably. We will make extensive use of TeXworks in this book.
- **Uninstall TeX Live**: Use this before you install a new TeX Live version from scratch, or if you would like to install MiKTeX instead.

Now, we will go through the offline installation of TeX Live.

Installing TeX Live offline

Every year, the TeX user group creates a TeX software collection DVD. You can request a copy here: `https://tug.org/dvd`.

You can also download it yourself. We will now download an ISO image of TeX Live with a size of about 6 GB. After extraction, we can burn it on a DVD and run the installation from there. To do this, follow these steps:

1. Visit the download area at `https://tug.org/texlive/acquire-iso.html`.
2. Download the `texlive.iso` file. If possible, use a download manager, especially if your internet connection is not stable.
3. Either burn the ISO file on a DVD using burning software that supports the ISO format, or extract it to your hard disk drive. For example, the free program **7-Zip** can extract ISO files.
4. Among the extracted files or on your DVD, you will find the `install-tl` and `install-tl-advanced` installer batch files. Choose one, start it, and go through the installation similar to the online installation. More information is available at `https://tug.org/texlive/quickinstall.html`.

Installing TeX offline was just like the first installation. Still, you've got all the data this time, and you won't need an internet connection during the installation or for another installation. This download approach is especially recommended if it's foreseeable that you will install TeX Live on another computer later, or if you would like to give it to friends or colleagues.

As TeX also runs on other operating systems, let's take a brief look at the other systems.

Installing TeX Live on other operating systems

TeX also runs on a lot of systems other than Windows. Here's a quick glance:

- **macOS**: You can download a customized version of TeX Live at `https://tug.org/mactex`. Download the huge `MacTeX.pkg` file and double-click on it to install. It will print very straightforward instructions. MacTeX includes **TeXShop**, an editor very similar to **TeX-works**. The latter can be separately installed.
- **Ubuntu Linux**: Use the **Software Center** to install TeX Live packages, or run `sudo apt-get install texlive-full` to get everything.
- **Debian Linux**: Use the **Synaptic** tool to install TeX Live packages, or run `apt-get install texlive-full` (via `sudo` or as root user) to get everything.
- **Red Hat, CentOS, and Fedora Linux**: Use the **Red Hat package manager**, or `yum` via Command Prompt, such as `yum install texlive-scheme-full`, or DNF: `sudo dnf install texlive-scheme-full`.
- **Others**: Visit `https://tug.org/texlive/quickinstall.html` and follow the instructions.

If you want the latest version of TeX Live, consider downloading and installing the most current version from the official website instead of the version from the operating system's repositories, as mentioned in the last point.

In the next section, we'll look at how to update your installation and add new packages.

Updating TeX Live and installing new packages

The LaTeX developers regularly release updates, both for new features and for bug fixes. It's a good idea to update your system from time to time.

To do this, open the **Start** menu, go to the **TeX Live** folder, and launch the **TeX Live Manager**, which is also referred to as `tlmgr` for short, and also called **TeX Live Shell**. This application is used for both updating and installing additional packages.

Take a look at the following screenshot so we can talk about how to use the TeX Live Manager:

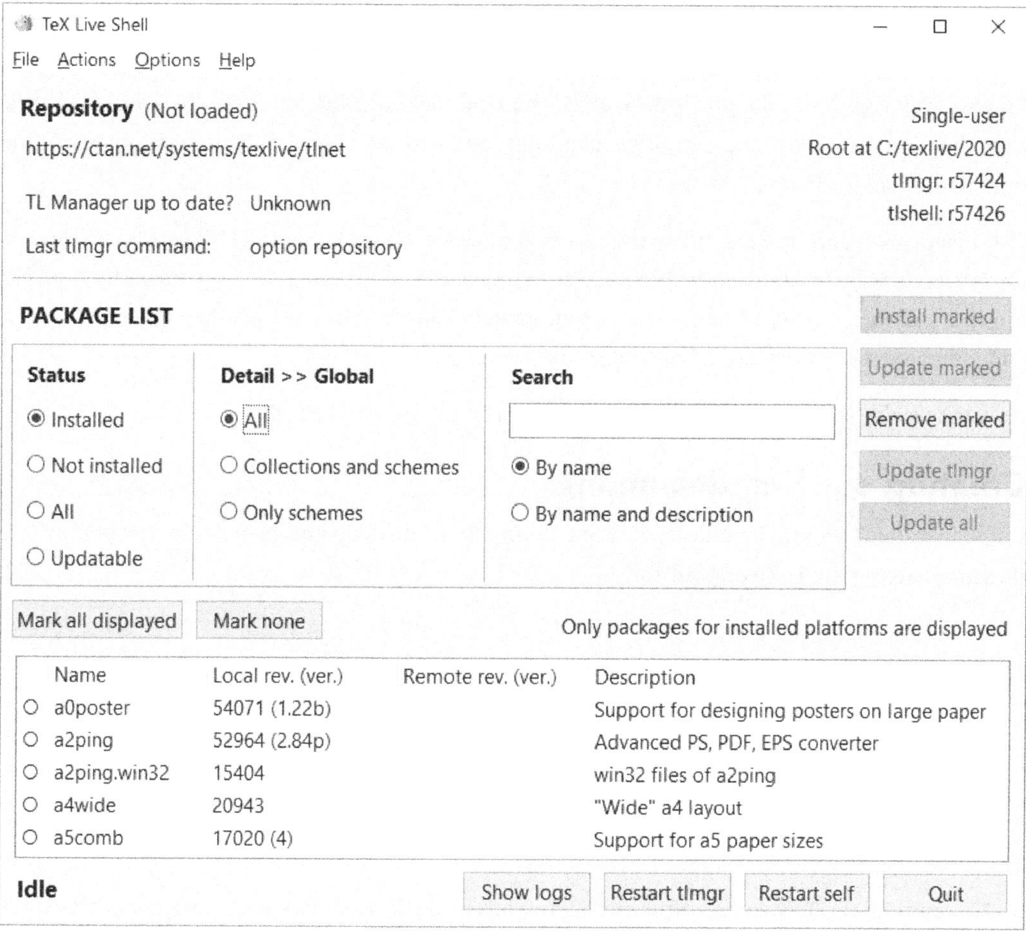

Figure 1.7 – The TeX Live Manager

The first section in the TeX Live Manager shows the **Repository** section. A **repository** is a server that hosts a TeX Live software archive. If the default repository is unavailable or is too slow in your area, you can click on **Options** to choose another repository from a list.

Click on **File | Load repository** to synchronize LaTeX with the latest software status.

The **PACKAGE LIST** section lets you search packages by name or filter the view to see all available packages or only those installed, not installed, or updatable. In the middle of *Figure 1.7*, you can see an option to adjust the granularity to see all the packages, all collections and schemes, or only schemes.

The lower section shows the matching packages when a filter is selected, with a short description and the version number. You can choose packages here. Then, you can click on **Install marked** if you would like to install the selected packages or click on **Remove marked** to uninstall them.

An easy way is just to click on **Update all**. If the **Update tlmgr** button is enabled and clickable, then there's a TeX Live Manager update available, and you can click the **Update tlmgr** button to update the tool itself.

The update procedure applies only to the current TeX Live version. A new TeX Live version is released every year, usually in March. It would be best to uninstall the current TeX Live version for a yearly upgrade and then install the new version from scratch. You can check the planned release schedule and estimated dates at `https://tug.org/texlive`.

Now that we've set everything up, it's time to start writing with LaTeX.

Creating our first document

We've installed TeX and an editor; now, let's jump in at the deep end by writing our first LaTeX document using the TeXworks editor.

Mac users, please use the *Cmd* key when you see the *Ctrl* key here.

Our first goal is to create a simple document that prints just one sentence. We want to use it to understand the basic structure of a LaTeX document. Follow these steps:

1. Launch the TeXworks editor by clicking on the desktop icon or by opening it in the **Start** menu. In *Figure 1.8*, you can see the editor with the menu, buttons, and toolbar.
2. Click the **New** button (or press *Ctrl* + *N*) or choose **File | New** in the menu.
3. Enter the following lines:

```
\documentclass{article}
\begin{document}
This is our first document.
\end{document}
```

4. Click on the **Save** button (or press *Ctrl* + *S*) to save the document. Then, choose a location to store your LaTeX documents, ideally in their own folder.

5. Make sure that in the drop-down field in the TeXworks toolbar, **pdfLaTeX** is selected (this should be the default anyway):

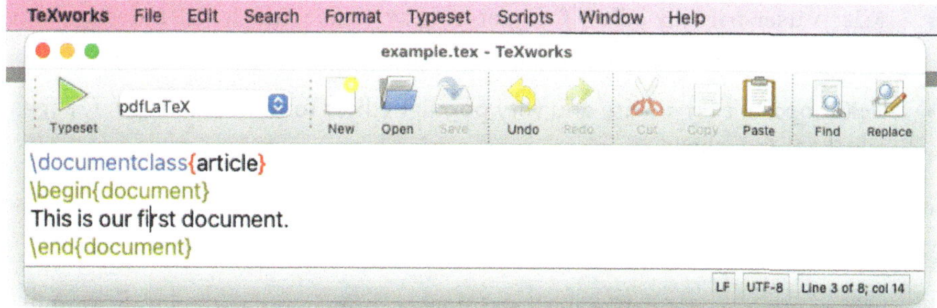

Figure 1.8 – The TeXworks editor

6. Click the **Typeset** button or press *Ctrl* + *T*.
7. The output window will automatically open. Have a look at it:

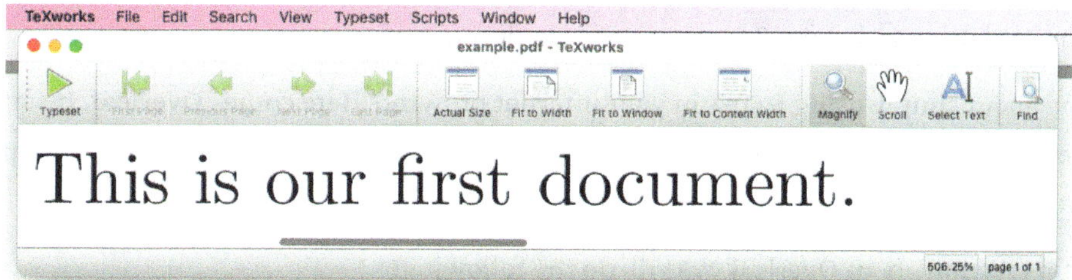

Figure 1.9 – The PDF output in the TeXworks editor

That covers the first few moments in the life of a LaTeX document. You'll edit, typeset, check the output, and edit again. Just remember to save your work frequently. Luckily, typesetting a document automatically saves it.

Unlike traditional word processors, LaTeX doesn't show formatting changes immediately, but you're always just one click away from seeing the final result.

Checking out advanced LaTeX editors

If you're comfortable with complex software and prefer a feature-rich, powerful editor, take a look at the following LaTeX editors. Visit their websites to find screenshots and to read about their features:

- **TeXmaker**: A cross-platform editor running on Windows, Linux, macOS, and other Unix systems: `https://xm1math.net/texmaker`

- **TeXstudio**: Another full-featured editor that runs on all major platforms: `https://texstudio.org`
- **Kile**: A user-friendly editor for operating systems with the KDE desktop environment, such as Linux: `https://kile.sourceforge.io`
- **TeXShop**: An easy-to-use and very popular editor for macOS: `https://pages.uoregon.edu/koch/texshop`

All of these editors are free and open source. You can find more options at `https://latexguide.org/editors`.

Online editors work on any device with an internet connection. Let's closely examine an online editor and compiler in the next section.

Working with LaTeX online

It is recommended that you install LaTeX on your computer, but it can take up about 8 GB on your hard drive and about 2 hours or more to do so.

How about simply using LaTeX in your internet browser? Here comes **Overleaf**. It's a purely online LaTeX service that mathematicians enthusiastic about TeX started in 2011. You can access it through this link: `https://www.overleaf.com`.

In this section about Overleaf, we will do the following:

- Check the Overleaf requirements
- Look at the benefits of Overleaf
- Evaluate possible caveats
- Use the Overleaf editor
- Try out Writefull

Let's go online now.

What Overleaf requires and delivers

To use Overleaf, you only need an internet browser, such as Firefox, Chrome, Opera, Edge, or Safari. You don't need any local software such as a LaTeX compiler, editor, or PDF viewer.

It is free for basic usage, and that includes *a lot*. It provides a complete TeX Live with unlimited projects, a feature-rich editor, real-time collaboration with another user, and hundreds of ready-to-use templates. However, bigger documents may time out due to limited computing resources. In particular, a free account is very limited in compiling time.

An advanced personal or professional subscription costs money and provides advanced features, such as the following:

- More collaborators per project
- Support for larger documents requiring more compiling time
- Document history (navigate back and forth between revisions)
- Advanced bibliography management (Mendeley, Zotero, and Papers)
- Built-in symbol palette
- GitHub and Dropbox integration
- Priority access to personal support

The advanced features go beyond regular LaTeX functionality. You might be eligible for full access through your institution; many universities and research organizations partner with Overleaf to offer premium plans to their members.

Benefits of Overleaf

Let's look at what you can gain when using Overleaf, compared to using a traditional LaTeX system locally on your computer. With Overleaf, you can do the following:

- Work from any device, such as a PC, laptop, tablet, or smartphone
- Use it on a locked-down work computer where you can't install software yourself
- Access your projects from anywhere (be it a private, work, or library computer) once you log in with your password
- If you invite somebody to work with you, both of you can edit and instantly see each other's changes, making collaboration easy
- Have an automatic real-time preview of the PDF result while you type
- Track changes with a built-in version history
- Annotate LaTeX source code with comments and replies to comments
- Work with the latest LaTeX system without installing an update

However, there are some trade-offs with Overleaf. Let's take a look at those next.

Caveats of working online

It's worth noting a few limitations:

- You'll need a constant internet connection.
- As your documents are stored online, you have to trust Overleaf with data security and privacy. See `https://www.overleaf.com/legal`.
- You're tied to Overleaf's platform; until it updates its LaTeX version, it could lag slightly behind the official TeX Live.
- The performance depends on their servers and your network connection, not just on your own computer's performance.
- Overleaf could be unavailable due to maintenance or a technical incident affecting its services. In this case, visit `https://status.overleaf.com` for current information.

Now, let's take a closer look at how Overleaf works.

Creating our first document online

We want to create our own space on Overleaf in two steps. Then, we will start our first LaTeX project:

1. Register for the service. Either click on **Register** on the Overleaf home page or go to `https://www.overleaf.com/register`. Enter your email address and choose a password.
2. Log in to Overleaf. Either click on **Login** on the front page or go to `https://www.overleaf.com/login`:

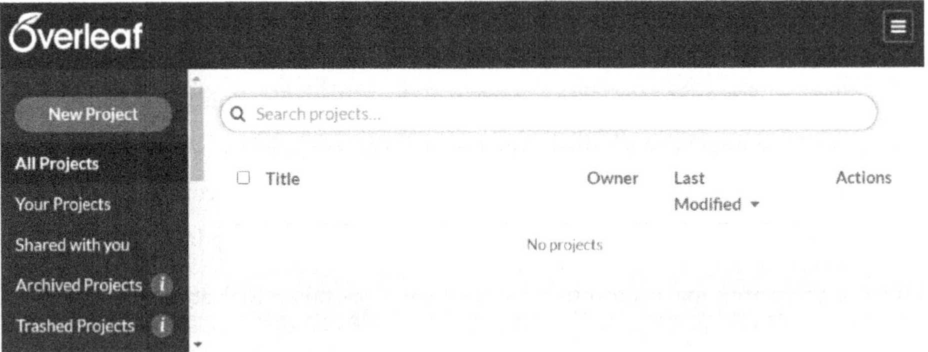

Figure 1.10 – Creating a new project

Why do we need to register?

To be reachable via an email address is basically to comply with data protection law. If you forget your password, you can ask Overleaf to send a password reset link to your email address. In general, by using your email address, you can prove your identity and ownership of your data if you ever need to.

3. Click the **New Project** button. A drop-down list will appear, where you can choose to have a blank project or one based on a template, such as a book, presentation, CV, or thesis template. For now, we just choose **Blank Project**.
4. Overleaf asks you for a name; choose one. That's it! This is what you have now:

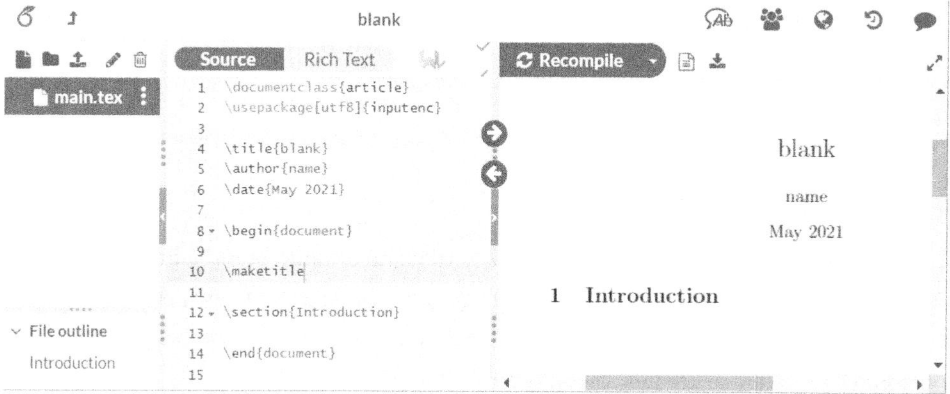

Figure 1.11 – A new project

It's not entirely blank as it contains a small code frame, so you get a quick start, and you can begin filling in your text.

The preview on the right side will be refreshed whenever you click the **Recompile** button or press *Ctrl + Enter*. You can enable automatic typesetting if you open the **Recompile** menu and choose **Auto Compile** in the drop-down menu, which is switched on in *Figure 1.12*.

The document refreshes frequently and automatically while you are typing:

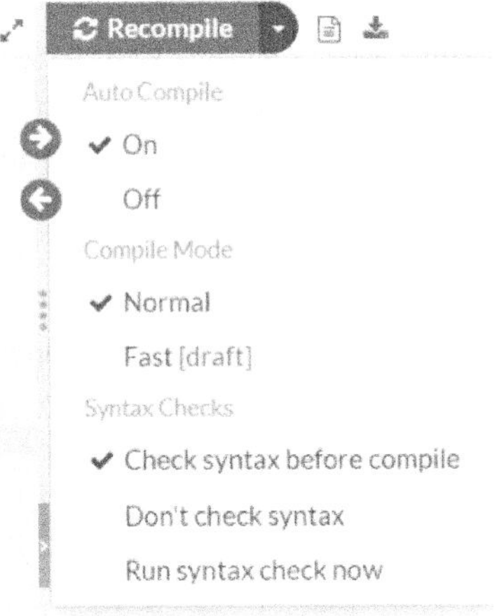

Figure 1.12 – Compiling settings

Since Overleaf is so different from classic LaTeX editors, let's have a closer look at it.

Exploring Overleaf

To quickly see a more complex document in action and to understand what you can expect from Overleaf, let's open the *Masters/Doctoral Thesis* template from `https://www.latextemplates.com/template/masters-doctoral-thesis`. Once you are there, simply click the **Open in Overleaf** button. Then, Overleaf creates a project for you.

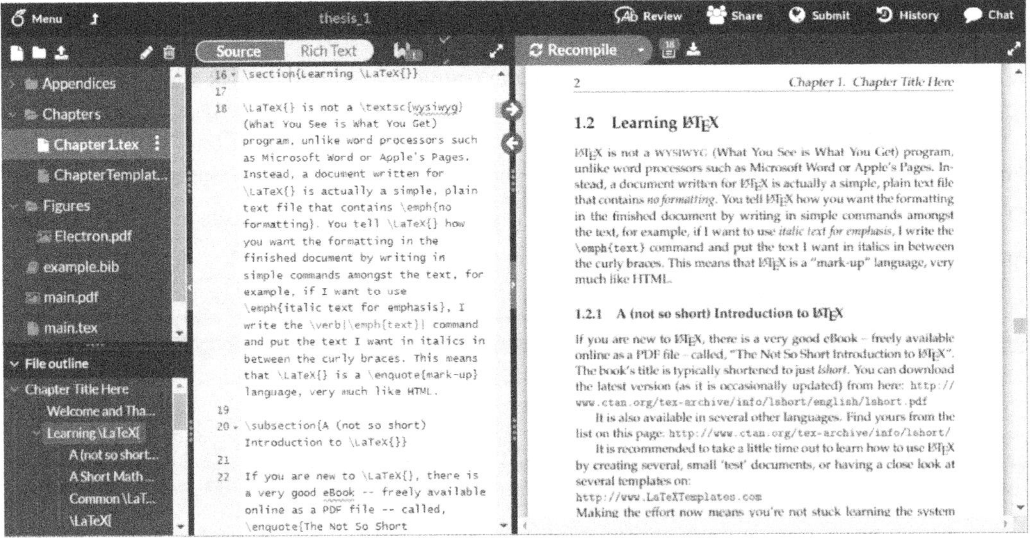

Figure 1.13 – Projects in the Overleaf editor

In *Figure 1.13*, on the very left, you can see your folder structure and files. Next to it, there will be the LaTeX source code. On the very right, you will see the output in a PDF preview, as in *Figure 1.11*.

In its most basic form, you write your code on the left, click the **Recompile** button, and see the result on the right, as we previously did in our example.

While you are working, Overleaf keeps track of the history. You can label versions to check them later. Clicking on the **History** button at the top shows you the versions:

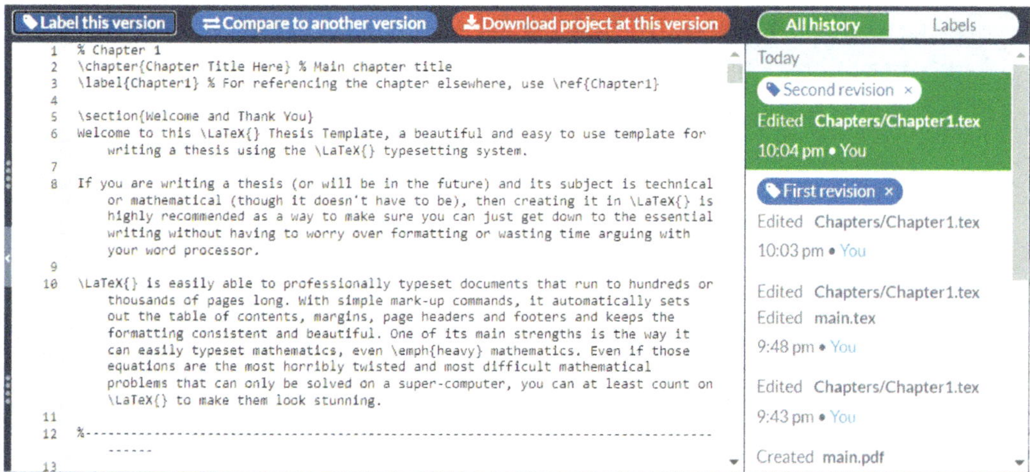

Figure 1.14 – Document history in Overleaf

Click a version on the right to switch to it.

Without looking at too many screenshots, this is what you can do if you click on the **Menu** button in the top-left corner:

- Let Overleaf count the words in your document, excluding code syntax such as commands and environments
- Synchronize with Dropbox or GitHub
- Choose the compiler (pdfLaTeX, classic LaTeX, XeLaTeX, or LuaLaTeX, for advanced users)
- Set a TeX Live version if you want to compile an old file with an old TeX Live version or switch to a newer one
- Select a main `.tex` document if your project consists of several documents
- Choose a visual style editor theme for the code markup and background, changing colors between light, pastel, dark, and others

- Choose the editor font (such as Consolas or Lucida) and the font size
- (De)activate spell-checking, auto-completion, auto-closing of brackets, and code-checking

The integrated spell-checker of Overleaf marks issues with a wavy line. Just right-click on it to get replacement suggestions, as shown in the following screenshot:

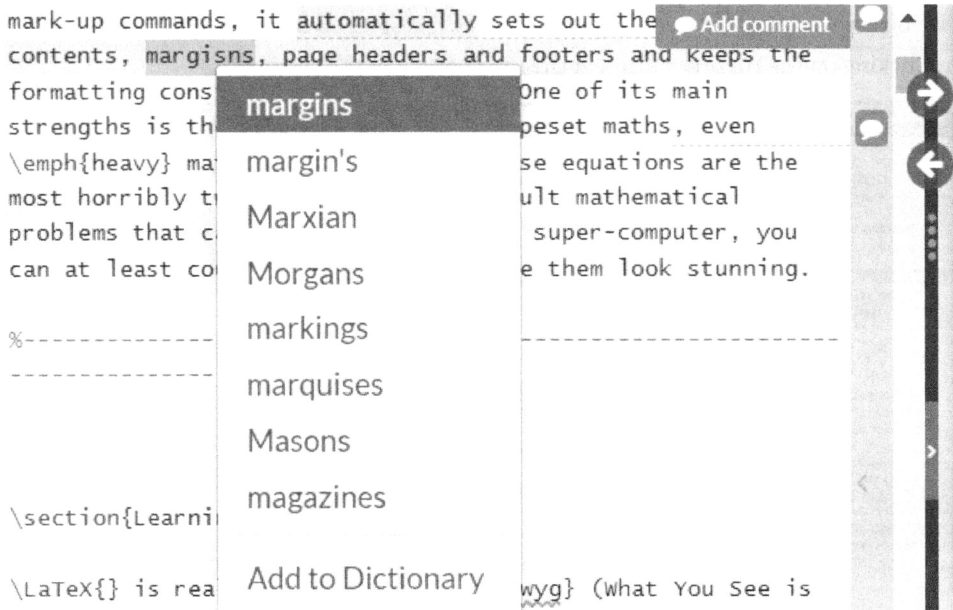

Figure 1.15 – Spell-checking in Overleaf

Apropos spell-checking, there's more.

Grammar and language feedback with Writefull

The **Writefull** Overleaf extension checks grammar and provides phrasing suggestions for your text. It's designed for scientific writing, and it's trained on millions of research journal articles. It can correct typos, grammar mistakes, vocabulary issues, punctuation, and more.

Let's see it in action on our previous thesis template example:

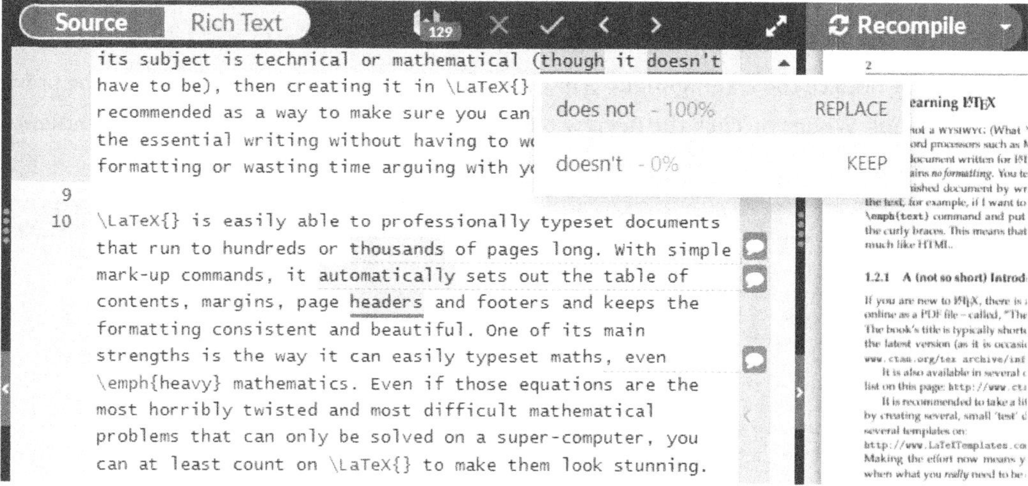

Figure 1.16 – Grammar checking with Writefull

While the spell-checker doesn't complain, the trained AI of the Writefull extension shows 129 potential issues and gives suggestions to the words underlined in red:

- `though` may be replaced by `although`
- `doesn't` may be replaced by `does not`, in formal writing

After `headers` (in the previous figure), there should be an Oxford comma, that is, a comma before the word and at the end of a list.

You can take it easy as I do in this book, not so formal, but your thesis or research articles and any scientific writing may benefit from these suggestions.

The Writefull extension was initially available for the Chrome browser. However, it was announced that other web browsers, such as Firefox, will be supported in the future. The basic version is free, and there is a premium version that you can read about on their website.

At the time of finishing this edition of the book, Writefull has been integrated into Overleaf itself. The full information is available at `https://docs.overleaf.com/integrations-and-add-ons/ai-features/writefull`.

Reviewing and commenting

You probably noticed the text highlighted in yellow and the speech bubble symbols in the previous screenshot. When you click the **Review** button, the review bar expands, and the comments will be shown:

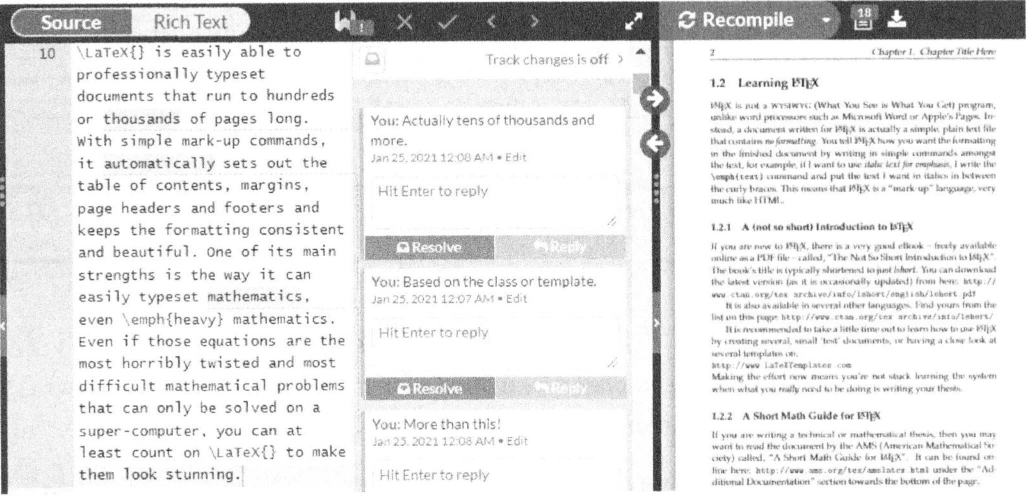

Figure 1.17 – Reviewing and commenting with Overleaf

You can mark text snippets, click on **Add comment**, write what you think, and also reply to other people's comments. That helps both in your own writing and in collaborating with colleagues or an editor.

This and the previously mentioned features should provide a good insight into current cloud LaTeX services.

The entire set of the *LaTeX Beginner's Guide* code examples can be opened in Overleaf simultaneously, with a single click. Visit `https://latexguide.org/code` to get the entire package and have your own project with all book examples included, ready to edit and compile.

The next section will provide you with a way to access additional supporting documentation and references while working with this book.

Accessing documentation

LaTeX offers hundreds of classes and packages, far more than any book could cover in full. Fortunately, most packages provide well-written documentation that you can easily access and read. As you work through this book, you'll find many useful packages. If you explore their documentation alongside the examples in the chapters, you'll be well on your way to becoming a skilled LaTeX user.

In the following chapters, you will learn about many LaTeX packages that provide additional capabilities. So, let's see how to access package documentation.

Once you've installed LaTeX, you can open a package manual directly on your computer after you have installed LaTeX:

- **On a Windows PC**: Open the **Start** menu, select the **TeX Live** folder, and click on **TeX Live command-line**. Alternatively, launch the Windows **cmd** (Command prompt) app.
- **On a Mac or any Linux computer**: Open the **Terminal** app.

Next, type `texdoc packagename` and press *Enter*. For example, typing `texdoc geometry` opens the PDF manual for the `geometry` package.

If you did not install LaTeX, you could obtain the documentation online. In your browser, open `https://texdoc.org/pkg/packagename`. That is just a template URL, so if you want the `geometry` package, you should type `https://texdoc.org/pkg/geometry`. That's an easy method for Overleaf users who don't have a locally installed documentation.

Throughout this book, you'll often see references to package manuals or documentation. Always remember that you can review documentation using the `texdoc` command or by visiting `https://texdoc.org`, as described earlier.

Finally, there's a lot of LaTeX-related documentation available online, including tutorials, guides, and reference materials. We will discuss these further in *Chapter 15, Using Online Resources.*

Summary

In this chapter, we explored the key benefits of LaTeX; soon, it will be our turn to use its virtues to achieve high-quality results. Furthermore, we covered installing, editing, and using LaTeX, both locally on your computer and online in the browser.

Now that you've got a functional and tested LaTeX setup in place, you're ready to write your own LaTeX documents. In the next chapter, we will talk about formatting text in detail.

2

Formatting Text and Creating Macros

In the last chapter, we installed LaTeX and wrote our first document using both the **TeXworks** editor and **Overleaf**. Now, we'll take a closer look at the structure of text, focusing on formatting and fine-tuning details.

In this chapter, we'll cover the following:

- Using logical formatting
- Understanding how LaTeX processes your input
- Modifying text fonts
- Using color
- Limiting paragraph width
- Breaking lines and paragraphs
- Changing text alignment
- Displaying quotations
- Creating commands
- Creating environments
- Using an alternative syntax

Through hands-on examples and trying the features, you'll learn some essential LaTeX concepts. By the end of this chapter, you'll be comfortable working with commands and environments and be able to define custom commands.

Remember that you can edit and compile all the examples online on the book's web page. This chapter's code can be found at `https://latexguide.org/chapter-02` and in the book's GitHub project.

As you begin working more extensively, you may occasionally encounter error messages. If that happens, refer to *Chapter 14*, *Troubleshooting*, for guidance and possible solutions.

Using logical formatting

Within a LaTeX document, we should avoid applying **physical formatting**, such as making words bold or italicized or choosing different font sizes. Instead, we should use **logical formatting**, such as declaring a title and the author, and then assigning a section heading. LaTeX takes care of the visual appearance, such as printing the title in a large font size and displaying section headings in bold.

In some examples later in this chapter, we will use physical formatting commands, such as making words bold or italic. However, that's for practicing font commands. The ultimate goal is to define our own logical commands using font commands.

In a good LaTeX document, physical formatting would only be used inside the definition of logical formatting commands. If you need a specific style, such as for highlighting keywords, you can define a logical command for it in the document preamble. Then, throughout the body of your document, you only use that logical command. This ensures consistent formatting throughout your text. Whenever you change your mind about formatting details, you can modify the logical commands in the preamble; no need to go through the entire document.

To get started, let's examine a brief example that illustrates the typical document structure.

Creating a document with a title and heading

Let's build a short example that uses some basic formatting. The document will include a title, the author's name, the date, a section heading, and some regular text.

Follow these steps:

1. Type the following code into your editor to start a small document:

   ```
   \documentclass[a4paper,11pt]{article}
   ```

2. Specify the title, author, and date:

```
\title{Example 2}
\author{My name}
\date{May 5, 2021}
```

3. Begin the document:

```
\begin{document}
```

4. Let LaTeX print the full title, which will include the author and date:

```
\maketitle
```

5. Add a section heading and write some text, then end the document:

```
\section{What's this?}
This is our second document. It contains a title and a section with
text.
\end{document}
```

6. Save the document by clicking the **Save** button (or pressing *Ctrl* + *S*). Give it a name, such as `example2.tex`.
7. Compile the document by clicking the **Typeset** button (or pressing *Ctrl* + *T*); this converts your code into a PDF file.
8. View the output:

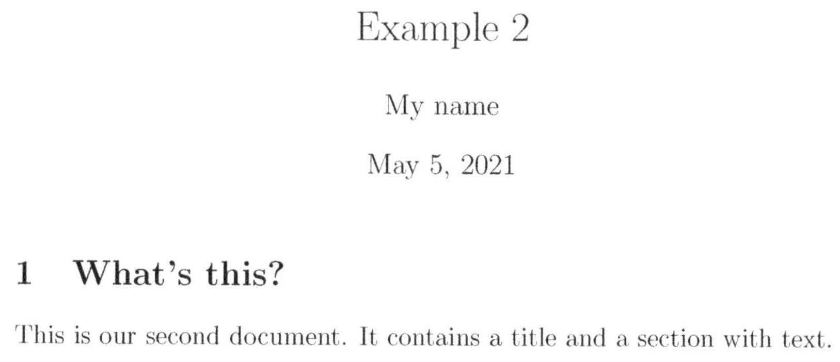

Example 2

My name

May 5, 2021

1 **What's this?**

This is our second document. It contains a title and a section with text.

Figure 2.1 – Text with a heading

In the TeXworks editor, the PDF preview appears automatically after you press the **Typeset** button. At the same time, a PDF file is created. In this case, it's called `example2.pdf` and stored in the same folder as your original code file, `example2.tex`.

In the first chapter, we talked about **logical formatting**; let's now look at this example with that in mind. Here's what we told LaTeX:

- The document type is `article`, and it is to be printed on A4 paper with a base font size of 11 points
- The document title is **Example 2**
- The document shows the author's name
- The document's date is set to **May 5, 2021**

Concerning the content of the document, we stated the following:

- The document begins with a title
- The first section is titled **What's this?**
- The section contains the following text: **This is our second document. It contains a title and a section with text.**

Notice that we didn't choose the font size of the title or section heading, or make anything bold or centered. LaTeX handles all of that formatting. Nevertheless, you're free to customize the way things look.

Once you've saved a document, there's no need to press the **Save** button again. TeXworks automatically saves it whenever you click the **Typeset** button.

Exploring the document structure

Let's take a closer look at the example we just created. A LaTeX document doesn't exist in isolation—typically, it's based on a reusable template called a **class**. A class provides an overall structure, along with customizable features, typically tailored to a specific purpose. There are classes for various types of materials, including books, journal articles, letters, presentations, and posters. You'll find hundreds of reliable classes in online archives and even on your computer after you've installed TeX Live. In our example, we chose the `article` class, a standard LaTeX class well suited for shorter documents.

The first line starts with `\documentclass`. This word begins with a backslash; such a word is called a **command** or a **macro**. We've already used a few such commands to specify the class and to state the document's metadata in our first example in this chapter: `\title`, `\author`, and `\date`. These commands store the properties; they don't produce visible output.

The first part of the document is called the **preamble**. This is where we choose the class, specify properties, and, in general, make document-wide definitions.

`\begin{document}` marks the end of the preamble and the beginning of the actual document. `\end{document}` marks the end of the document. Anything that follows would be ignored by LaTeX. Generally, such a piece of code framed by a `\begin` and `\end` command pair is called an **environment**.

Within the `document` environment, we used the `\maketitle` command to print the title, author, and date in a nicely formatted layout. With the `\section` command, we produced a heading, bigger and bolder than standard text. Finally, we added a short paragraph. To be clear: the preamble will never produce any output. Only what's in the document environment will be printed.

Now that you've seen how commands work, let's take a closer look at LaTeX's command syntax.

Using LaTeX commands

LaTeX commands begin with a backslash, followed by letters, usually lowercase, sometimes mixed with uppercase. They are usually named descriptively, with exceptions, though: later on, you will encounter commands consisting of just a backslash and a single special character.

Commands can take **parameters**. These are options that determine how the command performs its work. The values of parameters passed to commands are referred to as **arguments**. Arguments appear in curly braces when they are mandatory, or in square brackets when they are optional.

So, you'll see commands used in various ways, typically like one of these:

```
\command
\command{argument}
\command[optional argument]
\command[optional argument]{argument}
\command{argument}{another argument}
```

You see that some commands can take multiple arguments, with each of them enclosed in braces or brackets. Let's look closer at how it works.

If a command requires an argument by definition, that argument has to be provided; it's *mandatory*, denoted by curly braces. For example, calling `\documentclass` without specifying a class name would be pointless.

Commands can support arguments without strictly requiring them, so they are called **optional**. You may specify them, but it's not a must. If an optional argument isn't provided, the command will use a default one. Enclose such arguments in square brackets.

Let's look at the first example in *Chapter 1, Getting Started with LaTeX*. We wrote `\documentclass{article}` to choose the `article` class. Since we omitted an optional argument, this document was typeset with a base font size of 10 points because this is the class's default base font size. In the second document, we wrote `\documentclass[a4paper,11pt]{article}`; here, we overrode the default values, so now the document will be adjusted for A4 paper with a base font size of 11 points.

Most LaTeX commands, including those we define ourselves, consist of other commands. That's why LaTeX commands are also called **macros**, and the terms *macro* and *command* are used interchangeably. A command or macro that doesn't print something but just changes current settings, such as the font shape or text alignment, is also called a **declaration**.

There's a kind of command that applies changes only within a limited scope, such as a few lines or a code block. That's called an **environment**. Let's look at their syntax.

Using LaTeX environments

LaTeX environments are code blocks that start with `\begin` and end with `\end`. Both commands require the name of the environment as their argument.

A simple environment can look like this:

```
\begin{name}
    ...
\end{name}
```

Such environments can also be used for an existing declaration called `\name`, limiting the effect of the `\name` command.

Just like commands, environments can take arguments. Again, mandatory arguments go in curly braces while optional arguments go in square brackets. So, you may encounter an environment written like this:

```
\begin{name}{argument}
    ...
\end{name}
```

Or, you might see an additional optional argument:

```
\begin{name}[optional argument]{argument}
  ...
\end{name}
```

You can think of environments as declarations with a built-in scope. When `\begin` is called, the environment changes something particular, such as the layout, font, or other properties. That change stays in effect until an `\end` command is reached, at which point the change is canceled. The impact of the environment `name` is limited to the piece of code between `\begin{name}` and `\end{name}`.

Furthermore, the effect of all local declarations for formatting commands used within an environment will also end when the surrounding environment ends.

Now that we've covered the syntax of LaTeX commands and environments, let's look at how LaTeX understands our code.

Understanding how LaTeX processes your input

Before we continue writing, let's look at how LaTeX understands what we type:

- In addition to basic letters from the Latin alphabet, you can directly type (or copy and paste) accented characters, such as ä, ü, and ö, as well as characters from other languages, such as Greek or Cyrillic
- A single space in the input code appears as a space in the output document, while several consecutive spaces are treated as just one space
- A line break in the source code is treated like a space
- An empty line in the source code is treated as a paragraph break

Some characters have special meanings, like you saw before:

- A backslash, \, starts a LaTeX command or a LaTeX macro
- Curly braces and square brackets are used for command arguments
- A dollar sign, $, begins and ends math mode, which we'll explore in *Chapter 9, Writing Math Formulas*
- A percent sign, %, tells LaTeX to ignore the rest of the line

Let's expand on that last point: the percent sign begins a **comment**. Everything following a percent sign up to the end of the line will be ignored by LaTeX and won't be printed out. This allows you to insert notes into your document. It's often used in LaTeX templates to inform the user of what the template does or requires the user to do at a certain point. Note that the end of the line, normally behaving as a space, will also be ignored after a percent sign. That's the reason why you can often see a % sign at the end of lines in other people's code: to prevent the end of the line from acting like a space.

If you want to disable a command or a line of code temporarily, it's often better to insert a percent sign instead of deleting the command. That way, you're able to undo this change easily by just removing the percent sign.

If that is how the percent sign works, what should we do if we want to write something like 100% in our text? And what about the other special symbols? Let's figure out how to handle such cases in the next section.

Printing special symbols

Most of the time, you can simply type letters, digits, and punctuation directly into your LaTeX document. However, some characters are *reserved* for LaTeX commands and can't be used as is in your text. We've already seen such characters, such as the percent sign and curly braces. To include these characters in your output, LaTeX provides special commands to print such symbols.

Let's write a very short example printing out a dollar amount, a percentage, and a few other special symbols:

1. Create a new document and enter the following lines:

```
\documentclass{article}
\begin{document}
Statement \#1:
50\% of \$100 equals \$50.
More special symbols are \&, \_, \{ and \}.
\end{document}
```

2. Click the **Typeset** button to compile the document.
3. Check out the output:

Statement #1: 50% of $100 equals $50.
More special symbols are &, _, { and }.

Figure 2.2 – Special symbols

By placing a backslash in front of a special character, we turned it into a LaTeX command. The only purpose of this command is to print out that symbol.

You may be wondering how to print a backslash. To do this, use the `\textbackslash` command. Two consecutive backslashes, \\, are used as a shortcut for a line break. That might seem a bit unintuitive, but line breaks occur frequently, whereas backslashes are rarely printed in the output, so this shortcut has been chosen.

There's a huge range of additional symbols that we can use for math formulas, chess notation, zodiac signs, music scores, and more. We won't need to deal with those symbols for now, but we shall return to that subject in *Chapter 9*, *Writing Math Formulas*, when we'll need symbols to typeset math formulas.

Now that we know how to enter plain text, let's move on to formatting our text.

Modifying text fonts

LaTeX applies some formatting automatically; for example, we've seen that section headings are bigger than regular text and bold-faced. Now we will learn how to change the appearance of the text ourselves.

Adjusting the font shape

In this example, we'll emphasize an important word in our text, and we will see how to make words appear in **bold**, *italics*, or slanted. We'll also figure out how to highlight words inside a section of text that's already emphasized.

Let's take a look:

1. Create a new document containing the following code:

```
\documentclass{article}
\begin{document}
Text can be \emph{emphasized}.
Besides from \textit{italics}, words can be
```

```
\textbf{bold}, \textsl{slanted}, or typeset
in \textsc {Small Caps}.
Such commands can be \textit{\textbf{nested}}.
\emph{See how \emph{emphasizing} looks when nested.}
\end{document}
```

2. Click **Typeset** and have a look at the output:

   Text can be *emphasized.*
   Besides from *italics*, words can be **bold**, *slanted*, or typeset in SMALL CAPS.
   Such commands can be ***nested***.
   See how emphasizing *looks when nested.*

Figure 2.3 – Emphasizing phrases

At first, we used the `\emph` command, passing a word as its argument. This word will be typeset in italics because LaTeX uses italics to emphasize text.

Most text-formatting command names follow the pattern `\text**{argument}`, where ** stands for a two-letter abbreviation, such as `bf` for bold face, `it` for italics, and `sl` for slanted. The argument text will then be formatted accordingly. After the command, LaTeX continues typesetting subsequent text in the previously used style.

We nested the `\textit` and `\textbf` commands to combine styles, so the text appears in both italics and bold.

It's worth noting that most font commands will show the same effect if they are applied repeatedly, such as `\textbf{\textbf{words}}`. Here, the words will still be in the same bold style as with `\textbf{words}`, not bolder.

However, the `\emph` command behaves differently. We've seen that `\emph` changes text to italics, but if we use `\emph` on a piece of text that is already in italics, it will switch back from italic to upright font. Imagine a vital theorem completely typeset in italic, and you would like to highlight a word within this theorem. That word should not be in italics because it should be distinguishable from the surrounding text, so it's formatted in an upright font again. In other words, the `\emph` command toggles between upright and italics.

Use font changes with care. Combining font shapes, such as big, bold, and italic, at the same time, can be visually overwhelming and may be perceived as poor style. Use available font styles thoughtfully and apply them consistently throughout your document.

Choosing the font family

By default, LaTeX uses a **serif** font, also called a **Roman** font. Such fonts have small lines or strokes attached to letters, called **serifs**. Fonts without such serifs are **sans-serif** fonts, using the French word *sans*, meaning *without*.

Compare the two lines in *Figure 2.4*. Look closely at the first letter **T**; you'll clearly notice the difference between serif and sans-serif fonts:

This is serif
This is sans-serif

Figure 2.4 – Serif versus sans-serif font

These different kinds of fonts are referred to as **font families** or **typefaces**.

Another available typeface is **monospaced**, where all letters have the same width. Such fonts are also called **typewriter** fonts.

Let's switch font families in a short example document. We'll begin with bold text, but using bold text with serif can look very heavy. So, we will use sans-serif bold text instead. The following text will contain an internet address, and we will choose a typewriter font to set it apart.

Follow these instructions:

1. Create a LaTeX document with the following code:

```
\documentclass{article}
\begin{document}
\textsf{\textbf{Get help on the Internet}}
\texttt{https://latex.org} is a support forum for \LaTeX.
\end{document}
```

2. Click on **Typeset** and look at the result:

Get help on the Internet
`https://latex.org` is a support forum for LaTeX.

Figure 2.5 – Text with a URL

Here, we encountered further font commands. Using `\textsf`, we've chosen a sans-serif font in the heading line, and we applied the `\texttt` command to get a typewriter font for the internet address. These commands work just like the font commands we've seen earlier.

Serifs, those small decorative details at the ends of a letter's strokes, improve readability by guiding the reader's eyes along the line. That's why they are often preferred in printed books and newspapers.

Headings, however, are often set without serifs. Sans-serif fonts are also a good choice for screen text due to their better readability on lower-resolution displays or smaller screens, such as mobile phones. For this reason, sans-serif fonts are often preferred for text in e-books, on internet pages, and often for technical texts.

Monospaced (typewriter) fonts are ideal for displaying source code of computer programs, in printed documents as well as in text editors. As in our previous example, this book generally uses a typewriter font to distinguish internet addresses and source code from the surrounding text.

So far, we've seen commands that apply formatting to the text in the argument in curly braces. LaTeX also offers commands that don't take arguments and work like switches.

Using the following instructions, we'll modify the previous example, this time using font family switching commands:

1. Edit the previous example to get the following code:

   ```
   \documentclass{article}
   \begin{document}
   \sffamily\bfseries Get help on the Internet
   \normalfont\ttfamily https://latex.org\normalfont\ is
   a support forum for \LaTeX.
   \end{document}
   ```

2. Click on **Typeset** to compile.
3. Compare the output to the previous example; it looks the same.

We switched to the sans-serif typeface using the `\sffamily` command. The `\bfseries` command additionally switched the text to bold. We used the `\normalfont` command to return to the default LaTeX font, and then we applied the `\ttfamily` command to switch to a typewriter font. After the internet address, we used `\normalfont` again to switch back to the default font.

Such switching commands don't produce any visible output on their own. They just change how the following text is rendered. That's why they are also called **declarations**.

Let's summarize the font commands and their corresponding declarations along with their meanings:

Command	Declaration	Meaning
\textrm{...}	\rmfamily	roman family
\textsf{...}	\sffamily	sans-serif family
\texttt{...}	\ttfamily	typewriter family
\textbf{...}	\bfseries	**bold-face**
\textmd{...}	\mdseries	medium
\textit{...}	\itshape	*italic shape*
\textsl{...}	\slshape	*slanted shape*
\textsc{...}	\scshape	SMALL CAPS SHAPE
\textup{...}	\upshape	upright shape
\textnormal{...}	\normalfont	default font

Figure 2.6 – Font commands

The corresponding declaration to the `\emph` command is `\em`.

Limiting the effect of commands

In the previous example, we wrote `\normalfont` to switch the font back to LaTeX's default, but there's another way. We can use curly braces to tell LaTeX where to apply a command and where to stop it:

1. Shorten and modify our font shape example that produced *Figure 2.3* to get this code:

```
\documentclass{article}
\begin{document}
Besides from {\itshape italics}, words can be
{\bfseries bold}, {\slshape slanted}, or typeset
in {\scshape Small Caps}.
\end{document}
```

2. Click the **Typeset** button and check out the output:

Besides from *italics*, words can be **bold**, *slanted*, or typeset in SMALL CAPS.

Figure 2.7 – Using declarations to change the font weight and shape

When we wanted to apply a declaration to change the font, we started with an opening curly brace, {, and let the font declaration command follow. The effect of that command lasts until it's stopped by the corresponding closing brace, }.

An opening curly brace tells LaTeX to begin a **group**. The following commands are valid for the subsequent text until a closing curly brace ends the group. Groups can also be nested. Here's an example:

```
Normal text, {\sffamily sans serif text {\bfseries and bold}}.
```

The area where a command is valid is referred to as its **scope**. We have to be careful to complete each group. For every opening brace, there has to be a matching closing brace; otherwise, LaTeX will produce an error.

So, in short, groups are defined by curly braces and they contain and confine the effect of local commands.

Exploring font sizes

Let's try out all the font sizes available through LaTeX's built-in font size commands:

1. Create a document with the following code:

```
\documentclass{article}
\begin{document}
\tiny We \scriptsize start \footnotesize very
\smallsmall, \normalsize get \large big \Large
and \LARGE bigger, \huge huge, \Huge gigantic!
\end{document}
```

2. Click **Typeset** and observe the output:

We start very small, get big and bigger, huge, gigantic!

Figure 2.8 – Font sizes

We used all 10 available font size declarations, starting small with `\tiny` and ending with `\Huge`. There are no corresponding commands that take arguments, so if you want to limit their scope, you would have to use curly braces, just as we learned earlier.

The actual font size scales with the base font size of your document. So, if your document has a base font size of 12 points, then `\tiny` would result in text bigger than with a base font of 10 points.

Use the `\footnotesize` command if you want the same font size as LaTeX uses for footnotes; use `\scriptsize` for a size that matches LaTeX's subscripts and superscripts. Document classes provide carefully selected and well-suited font size steps, so you usually don't need to set a specific physical size.

If you want fine-grained control or advanced tweaks, these are covered in *Chapter 3, Adjusting Fonts*, in the *LaTeX Cookbook*, published by *Packt Publishing*.

Using color

While the font style can help differentiate text, a bit of color can make content stand out more clearly, for emphasis, highlights, or getting attention. LaTeX provides the basic `color` package, but we will use the more versatile `xcolor` package right away.

We begin with a brief example, highlighting a short line of text in bold red to emphasize its importance. Follow these instructions:

1. Create a document with the following code:

```
\documentclass{article}
\usepackage{xcolor}
\begin{document}
\textcolor{red}{\textbf{Important:}}
  Always keep backup copies of your documents.
\end{document}
```

2. Click **Typeset** and look at the result:

Important: Always keep backup copies of your documents.

Figure 2.9 – Text with color

The `\textcolor` command takes two mandatory arguments: the color name and the text to be colored. In our example, we wrote `\textcolor{red}{...}` to color the word **Important:** in red. We also made the word bold using `\textbf`, showing how to combine two formatting commands. You can use standard colors such as red, blue, green, yellow, black, and even white. The `xcolor` package provides hundreds of predefined named colors, such as `LimeGreen` or `DarkCyan`, with huge tables in its manual. You can open that manual by running `texdoc xcolor` at the command line, or visit `https://texdoc.org/pkg/xcolor`.

As with font-switching commands, LaTeX also provides a declaration-style version for setting the text color across larger portions of text: the `\color` command acts as a switch; once issued, it affects all subsequent text until the end of the current group or environment. This code would result in the same output as the previous figure:

```
{\color{red}\bfseries Important:}
```

The `xcolor` package lets you easily mix colors using a simple syntax:

```
name1!percent1!name2!percent2!...
```

This means: take `percent1` of `name1`, `percent2` of `name2`, and so on. The total should ideally sum to 100%, but if the last percentage is missing, `xcolor` just fills in the rest. Here are a few examples to show how it works:

- For a light gray, you can write `\color{black!20}` to get 20% blackness
- To blend 30% red with 70% yellow, use `\color{red!30!yellow}`
- To combine 50% red, 20% green, and the remaining 30% of the mix shall be blue, write `\color{red!50!green!20!blue}`

I usually take this easy way to mix the base colors until it looks how I want. You can define your own color names, and you can use different color models for specifying numeric values such as **RGB**, **CMYK**, or even **HTML** codes, which is explained in the manual over 60 pages; but we'll keep it at this.

Now that we have seen how to format words and phrases, let's move on to formatting entire paragraphs.

Limiting paragraph width

In most cases, we write text just from left to right across the entire text area. But sometimes you may want to narrow the width of a paragraph, for instance, when placing text beside a picture.

In the following sections, we'll look at how to work with **paragraph boxes** in LaTeX.

Creating a narrow text box

In this example, we'll explain the acronym **TUG** in a text column that's only 3 cm wide. Here's how:

1. Create a new document containing these four lines:

   ```
   \documentclass{article}
   \begin{document}
   \parbox{3cm}{TUG is an acronym. It means
   \TeX\ Users Group.}
   \end{document}
   ```

2. Click **Typeset** and review the output. It works, but we get loose lines:

TUG is an acronym. It means TEX Users Group.

Figure 2.10 – A narrow justified paragraph with big gaps

In this example, we used the `\parbox` command to create a column of fixed width. The first argument specifies the width, and the second argument contains the text.

`\parbox` formats the argument text to fit the specified width. By default, the text is fully justified; however, our example shows an obvious problem: keeping full justification could lead to awkward gaps in the text. There are a few options to improve the appearance:

- Introduce hyphenation, for example, breaking long words such as *acronym*
- Improve general justification behavior
- Disable full justification and switch to ragged-right alignment

We'll examine these techniques in the *Breaking lines and paragraphs* and *Changing text alignment* sections later in this chapter.

But first, let's take a closer look at how `\parbox` actually works.

Producing paragraph boxes

In most cases, we simply need a text box with a certain width. However, sometimes, we also want to control how it aligns with the surrounding text. The `\parbox` command supports an optional argument for alignment. The general form of the `\parbox` command looks like this:

```
\parbox[alignment]{width}{text}
```

These are the meanings of the parameters:

- `alignment` is the optional argument for vertical alignment with the surrounding text:
    - `t` aligns at the baseline of the top line of the box.
    - `b` aligns at the baseline of its bottom line.
    - `c` aligns its vertical center with the center of the adjacent text line. That's the default so that you can omit it.

- `width` sets the width of the box. You can use standard length units without spaces, such as `3cm`, `44mm`, or `2in`.
- `text` is the content you want to put in that box. It should be a short paragraph or phrase. For more complex or multi-paragraph content, consider using the `minipage` environment instead. We will look at this in the next section.

Here's a demonstration of the effect of the alignment parameters:

```
\documentclass{article}
\begin{document}
Text line
\quad\parbox[b]{1.8cm}{this parbox is aligned at its bottom line}
\quad\parbox{1.5cm}{center-aligned parbox}
\quad\parbox[t]{2cm}{another parbox aligned at its top line}
\end{document}
```

The `\quad` command produces some space; we used it to separate the boxes a bit. Here's the output:

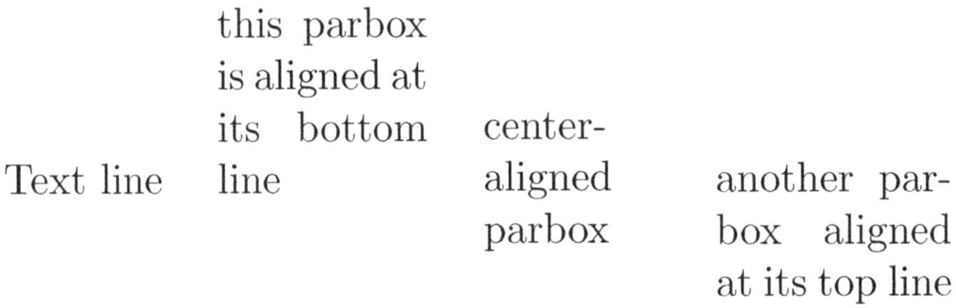

Figure 2.11 – Aligned paragraph boxes

Figure 2.11 shows how the alignment arguments work. Using the preceding text as the baseline, the following boxes are aligned, respectively, at the bottom line, the top line, or the center.

Exploring further paragraph box features

The `\parbox` command offers even more control. If you need advanced positioning, here's the complete syntax:

```
\parbox[alignment][height][inner alignment]{width}{text}
```

The meanings of the new arguments are as follows:

- `height`: By default, the box will have just the natural height of the text inside. Use this argument if you want to change the height of the box to make it bigger or smaller than the content.
- `inner alignment`: If the height of the box is different from the natural height of the contained text, you might want to adjust the text position. You can specify the following values:
  - `c`: Center the text in the box vertically
  - `t`: Align text to the top of the box
  - `b`: Align text to the bottom
  - `s`: Stretch the text vertically to fill the box (if possible)

  If you omit this argument, the `alignment` argument will be used here as the default value.

Take our previous demonstration example and try the effect of the optional arguments. To better see the effect, use the `\fbox` command: if you write `\fbox{\parbox[...]{...}{text}}`, the entire `parbox` will get a frame.

Using minipages

While `\parbox` works well for small amounts of text, it's hard to manage longer content. A closing brace could easily be forgotten or overlooked. The `minipage` environment would then be a better choice. A minipage is a small area on your main page that LaTeX keeps together without breaking across pages.

In this example, we will use the `minipage` environment instead of `\parbox` to get a sample of text with a width of just 3 cm:

1. Modify the `parbox` example to get the following code:

```
\documentclass{article}
\begin{document}
\begin{minipage}{3cm}
TUG is an acronym. It means \TeX\ Users Group.
\end{minipage}
\end{document}
```

2. Click on **Typeset** and look at the output:

TUG is an acronym. It means TEX Users Group.

Figure 2.12 – A minipage example

By using `\begin{minipage}`, we started a "page within a page." We specified the width of 3 cm as a mandatory argument. From that point onward, LaTeX will wrap text lines to fit within the 3 cm width; they will be automatically wrapped and fully justified. We ended this restriction with `\end{minipage}`. Any following text would run over the complete body text width.

A `minipage` environment will never have a page break within, so that's a way to prevent page breaks for content that you want to keep together. If the content of a `minipage` environment doesn't fit on the current page, LaTeX will move it entirely to the next page.

The `minipage` environment accepts the same arguments as `\parbox` with the same meanings.

When text is wrapped in a box or simply on a regular line, LaTeX will automatically fit it to the page layout reasonably well. However, we may still want to fine-tune line breaking and justification. Let's see how to do that in the following sections.

Breaking lines and paragraphs

In most cases, you don't need to worry about how lines break while you're writing. Simply type your text in your editor, and LaTeX will format it to fit the line and handle the justification. To start a new paragraph, simply insert an empty line before you continue with your text for the following paragraph.

Now, we will discuss how to control line wrapping. First, we'll examine ways to improve automatic hyphenation, and second, we'll learn commands to insert breaks directly.

Improving hyphenation

If you look at longer documents, you'll often notice how impressively LaTeX justifies paragraphs and distributes spacing between words evenly. If necessary, LaTeX will break words with hyphens at the end of the line to improve alignment.

LaTeX already uses excellent and very sophisticated algorithms for hyphenation, but in rare cases it can't find an acceptable way to break a word. The previous example pointed out this problem: breaking the word *acronym* would improve the output, but LaTeX does not know where to divide it. We shall find out how to solve that.

No matter how reasonable the justification is, text in very narrow columns is extremely hard to justify. To achieve full justification, LaTeX may insert significant gaps between the words, but we want to avoid that.

In the following example, we'll tell LaTeX how a word could be divided, to give LaTeX more flexibility for paragraph justification:

1. Insert the following line into the preamble of the previous example:

   ```
   \hyphenation{acro-nym}
   ```

2. Click on **Typeset** and look at the output:

   TUG is an acro-
   nym. It means TEX
   Users Group.

Figure 2.13 – A paragraph with improved hyphenation

We've instructed LaTeX that the word *acronym* may be split between *acro* and *nym*. That allows LaTeX to insert a hyphen after *acro* at the end of the line and move *nym* down to the next line.

The `\hyphenation` command tells LaTeX where a word may be divided if needed. Its argument can contain several words separated by spaces. For each word, we can indicate multiple hyphenation points. For instance, we could extend the argument by more hyphenation points and more word variants, like this:

```
\hyphenation{ac-ro-nym ac-ro-nym-ic a-cro-nym-i-cal-ly}
```

You could also indicate division points within the body text by inserting a backslash followed by a hyphen, such as `ac\-ro\-nym`. However, by using the `\hyphenation` command in the preamble, you can collect all rules there, ensuring they are used consistently. Therefore, use it mainly in rare cases where LaTeX's automation fails.

Preventing hyphenation

Sometimes, you may want to prevent LaTeX from hyphenating a particular word. There are a few ways:

- Declare it in the preamble using it in the `\hyphenation` argument without any hyphenation points, such as `\hyphenation{indivisible}`
- Wrap the word in an `\mbox` command, which keeps the entire word together, like this: `The following word is \mbox{indivisible}`

Loading the `hyphenat` package gives us more control:

- `\usepackage[none]{hyphenat}` disables hyphenation entirely
- `\usepackage[htt]{hyphenat}` enables hyphenation for typewriter text; otherwise, such monospaced words won't be hyphenated by default, as it makes sense for URLs and code keywords

These optional arguments to `\usepackage` are called **package options**. They configure the behavior of a package. The mentioned options may be combined, separated by commas, like in `\usepackage[htt,none]{hyphenat}`. Even if you don't use the `none` option, you can disable hyphenation for short pieces of text using the `\nohyphens{text}` command. Try out these features if you want to benefit from them. The package documentation at `https://texdoc.org/pkg/hyphenat` provides more information on additional features that you may need, such as hyphenation after special characters such as numerals and punctuation.

Improving the justification

Today's most popular TeX compiler is **pdfTeX**, which outputs documents directly as a PDF. When Hàn Thế Thành developed pdfTeX, he added micro-typographic enhancements, subtle adjustments that improve the overall appearance. We can take advantage of these new features by using the `microtype` package.

Let's improve our previous example using the `microtype` package:

1. Insert the following line into the preamble of the previous example:

   ```
   \usepackage{microtype}
   ```

2. Click on **Typeset** and look at the output:

TUG is an acronym.
It means TEX Users
Group.

Figure 2.14 – A paragraph with better justification

What happened? By simply loading the `microtype` package without any options, we activated its default features. It introduces font expansion to tweak the justification and uses hanging punctuation to improve the visual appearance of the margins. This may reduce the need for hyphenation and avoid getting awkward gaps between words to achieve full justification. You've seen its effect on a narrow column, so imagine the improvement on wide text—keep that in mind and try it out later!

Though `microtype` provides powerful features and options for the advanced typesetter, we usually won't need to do more than just load it to benefit from it. There's extensive package documentation if you want to study it in depth: run the `texdoc microtype` command at the command line or open it at `https://texdoc.org/pkg/microtype`. `microtype` does nice tweaking, but it's not a cure-all; we should still take care of proper hyphenation when necessary.

Breaking lines manually

We can decide to end a line ourselves, overriding the automation. Now, we will learn about several commands with different effects for ending a line.

Let's typeset the opening of the famous poem *Annabel Lee* by Edgar Allan Poe. Since the poet determined exactly where each line has to end, we'll insert breaks manually.

So, let's write up the beginning of the poem:

1. Create a document containing these lines:

```
\documentclass{article}
\begin{document}
\noindent\emph{Annabel Lee}\\
It was many and many a year ago,\\
In a kingdom by the sea,\\
That a maiden there lived whom you may know\\
By the name of Annabel Lee
\end{document}
```

2. Click **Typeset** and review the output:

Annabel Lee
It was many and many a year ago,
In a kingdom by the sea,
That a maiden there lived whom you may know
By the name of Annabel Lee

Figure 2.15 – Manually broken lines

The very short \\ command ended the current line, and the following text was moved to the next line. However, it's different from a paragraph break as we're staying in the same paragraph. The `\newline` command is an alternative and has the same effect.

By default, LaTeX indents the first line in the paragraph, which is a useful convention for visually separating paragraphs. We used the `\noindent` command to suppress the paragraph indentation. You won't need it often because, after headings, there's no paragraph indentation by default. For removing paragraph indentation for the entire document, and using vertical inter-paragraph spacing instead, load the `parskip` package. You can see this in *Figure 2.23* and the corresponding code.

Note that although we inserted line endings, the text remained a single paragraph. Therefore, our line breaks didn't cause a paragraph indentation, as it's logically still part of the same paragraph.

Exploring line-breaking options

The \\ command supports the following optional arguments for fine-tuning:

- `\\[value]` inserts additional vertical space after the break with the requested `value`, such as `\\[3mm]`
- `\\*[value]` does the same but prevents a page break from occurring immediately afterward

Try, for example, changing the opening of our poem example to the following:

```
\emph{Annabel Lee}\\[3mm]
```

That inserts an additional 3 mm space between this line and the poem fragment, separating it as a heading.

There's another command called `\linebreak` that tells LaTeX to end the line while maintaining full justification. If needed, LaTeX will stretch the space between words to reach the right margin. This could cause unpleasant gaps—that's why that command is rarely used, or at least with an optional argument; `\linebreak[number]` can be used to tweak the line break. If the `number` is `0`, a line break is allowed; `1` means it's desired; `2` and `3` mark more insistent requests; and `4` will enforce it. The latter is the default behavior if no number is given.

Preventing line breaks

The `\linebreak` command has a direct opposite: `\nolinebreak`. This command prevents a line break at the current position.

Like its counterpart, it takes an optional argument to control the strength of the request. If you write `\nolinebreak[0]`, you merely suggest not to break the line here. Using `1`, `2`, or even `3` insists more and `\nolinebreak[4]` strictly forbids it. LaTeX presumes the latter option if you don't provide an argument.

The already-mentioned `\mbox{text}` command not only disables the hyphenation of a word, but you can also use it to prevent a line break within the enclosed `text`.

Instead of `\nolinebreak`, you can use the `~` shortcut, which stands for an inter-word space where no line break is allowed. For example, if you wrote `Dr.~Watson`, LaTeX would never let the title `Dr.` dangle lonely at the end of a line.

By default, LaTeX fully justifies the text, stretching lines to align with the right margin. LaTeX has an excellent method to adjust spacing for whole paragraphs, but sometimes it may result in undesirable big gaps between words on a justified line. Let's see how to disable this if we want to.

Changing text alignment

While your text will usually look well aligned if full justification is used, there may be occasions when it's not ideal. For instance, full justification could be displeasing when text lines are short; in such a case, it could be sufficient to align only at the left side. Let's see how to put this into practice, as well as how to right-align and create centered lines.

Creating ragged-right text

Recall our first `parbox` example, where full justification led to noticeable gaps between words. This time, we'll turn off right-margin justification to avoid such gaps:

1. Create a new document with the following lines:

   ```
   \documentclass{article}
   \begin{document}
   \parbox{3cm}{\raggedright
   ```

```
    TUG is an acronym. It means \TeX\ Users Group.}
\end{document}
```

2. Click on **Typeset** and observe the result:

TUG is an
acronym. It means
TEX Users Group.

Figure 2.16 – Left-justified text

We inserted the `\raggedright` declaration. From this point onward, the text will be ragged-right. In other words, the text will be moved to the left margin—"flushed-left"—leaving the right edge uneven. There won't be hyphenation.

Since we used this declaration inside a box, it only affects the contents within, like inside environments. After the box, the text will be fully justified again.

If you want the whole document to be ragged-right, you can use `\raggedright` in the preamble.

Creating ragged-left text

In some cases, we may want to achieve the opposite effect: flushing the text to the right margin instead. We can do this similarly by inserting the `\raggedleft` declaration. You can control line breaks using the \\ shortcut.

Centering text

To align text horizontally in the center of the page, we can use the `\centering` declaration. Let's do this with a few example lines.

We'll manually create a nice-looking title for our document; it should contain the title, the author, and the date, all of which will be centered:

1. Write a document containing this code:

```
\documentclass{article}
\pagestyle{empty}
\begin{document}
{\centering
    \huge\bfseries Centered text \\
    \Large\normalfont written by me \\
```

```
    \normalsize\today
}
\end{document}
```

2. Click **Typeset** to see the output:

Centered text

written by me

December 31, 2025

Figure 2.17 – Centered text

Because only the title should be centered, we used a group to restrict the effect of the `\centering` declaration. All text within the group will be horizontally aligned to the center. We also inserted a paragraph break with an empty line; it's recommended to do this before ending the group to apply our centering to the full paragraph. By using the closing brace, we ended the group. Any text placed after the closing brace will revert to the default alignment, not centered.

`\centering` is often used for pictures and tables, and on title pages.

Using environments for justification

LaTeX provides the `center` environment that centers text and adds vertical space around it.

Let's test it. We will reuse the excerpt of the Edgar Allan Poe poem again. This time, we'll center all lines:

1. Start a new document:

   ```
   \documentclass{article}
   ```

2. Now, load the `url` package so that we can also print a hyperlink at the end:

   ```
   \usepackage{url}
   ```

3. Start the document with an introductory line:

   ```
   \begin{document}
   \noindent This is the beginning of a poem
   by Edgar Allan Poe:
   ```

4. Now, write text in a `center` environment:

   ```
   \begin{center}
   ```

```
    \emph{Annabel Lee}
\end{center}
```

5. Again, write text for the body of the poem:

```
\begin{center}
    It was many and many a year ago,\\
    In a kingdom by the sea,\\
    That a maiden there lived whom you may know\\
    By the name of Annabel Lee
\end{center}
```

6. Add some text, including a URL pointing to the poem on the internet, and finish:

```
The complete poem can be read on
\url{http://www.online-literature.com/poe/576/}.
\end{document}
```

7. Click **Typeset** and see the result:

This is the beginning of a poem by Edgar Allan Poe:

Annabel Lee

It was many and many a year ago,
In a kingdom by the sea,
That a maiden there lived whom you may know
By the name of Annabel Lee

The complete poem can be read on `http://www.online-literature.com/poe/576/`.

Figure 2.18 – A centered poem within text

We began with `\noindent` again, avoiding the paragraph indentation. `\begin{center}` started the `center` environment, which begins a new paragraph, with added vertical spacing. `\end{center}` ended this environment. We used the `center` environment a second time, where we inserted `\\` to end the verses. After the `center` environment ends, some space follows, and the next paragraph begins at the left margin.

The corresponding environment for ragged-right text is called `flushleft`; here, LaTeX pushes everything within the environment to the left and leaves the right side ragged, and, similarly, for ragged-left text, it's the `flushright` environment.

Centering is a popular technique for emphasizing text. Another way is to indent it a bit and to add some vertical space before and after the text. It's commonly used to display a quotation. Let's see how to do that next.

Displaying quotations

When including a quotation from another author, it might be hard to see that it's a quote when it's directly embedded within the text. A common and effective practice to improve readability is to visually set off the quotation by indenting it on both the left and right sides. Let's illustrate this with quotes from famous physicists in the following example:

1. Create a new document with some introductory text:

```
\documentclass{article}
\begin{document}
\noindent Niels Bohr said: ``An expert is a person
who has made all the mistakes that can be made in
a very narrow field.''
Albert Einstein said:
```

2. Display the quotation using the `quote` environment:

```
\begin{quote}
    Anyone who has never made a mistake has never
    tried anything new.
\end{quote}
```

3. Add a concluding sentence, and finish:

```
Errors are inevitable. So, let's be brave
trying something new.
\end{document}
```

4. Click **Typeset** to compile and view the result:

Niels Bohr said: "An expert is a person who has made all the mistakes that can be made in a very narrow field." Albert Einstein said:

> Anyone who has never made a mistake has never tried anything new.

Errors are inevitable. So, let's be brave trying something new.

Figure 2.19 – Quotes

Firstly, we quoted *inline*, which is within the text flow of the paragraph. The ` shortcut produces a left quotation mark, also called a **backtick**, and the ' shortcut gives a right quotation mark. To get double quotes, we just typed two such characters. We call this **inline quoting**.

Then, we used the quote environment to display a quotation separated from the surrounding text. We did not begin a new paragraph for it, because the quotation is already set off in its own paragraph. That's called **displayed quoting**.

Quoting longer text

For short quotations, the quote environment works well. However, when you're quoting a more extended text passage that spans multiple paragraphs, you may prefer to keep the default paragraph indentation like in your surrounding text. The quotation environment will do this for you.

Let's quote a few benefits of using TeX and LaTeX, as described on a web page on CTAN:

1. Start a new document with this text:

```
\documentclass{article}
\usepackage{url}
\begin{document}
The authors of the CTAN team listed ten good reasons
for using \TeX. Among them are:
\begin{quotation}
  \TeX\ has the best output. What you end with,
  the symbols on the page, is as useable, and beautiful,
  as a non-professional can produce.
  \TeX\ knows typesetting. As those plain text samples
  show, TeX's has more sophisticated typographical
  algorithms such as those for making paragraphs
  and for hyphenating.
  \TeX\ is fast. On today's machines \TeX\ is very fast.
  It is easy on memory and disk space, too.
  \TeX\ is stable. It is in wide use, with a long
  history. It has been tested by millions of users,
  on demanding input.
  It will never eat your document. Never.
\end{quotation}
The original text can be found on
\url{https://www.ctan.org/what_is_tex.html}.
\end{document}
```

2. Click **Typeset** and look at the output:

    The authors of the CTAN team listed ten good reasons for using TEX. Among them are:

    > TEX has the best output. What you end with, the symbols on the page, is as useable, and beautiful, as a non-professional can produce.
    > TEX knows typesetting. As those plain text samples show, TeX's has more sophisticated typographical algorithms such as those for making paragraphs and for hyphenating.
    > TEX is fast. On today's machines TEX is very fast. It is easy on memory and disk space, too.
    > TEX is stable. It is in wide use, with a long history. It has been tested by millions of users, on demanding input. It will never eat your document. Never.

    The original text can be found on `https://www.ctan.org/what_is_tex.html`.

Figure 2.20 – A long section of quoted text

This time, we used the `quotation` environment to display multiple paragraphs. Like in regular text, blank lines separate the paragraphs; they are left-indented at their beginning just like in all our body text.

But what if we prefer not to use that paragraph indentation? Let's check out an alternative.

In this example, we'll avoid paragraph indentation and instead separate the paragraphs with some vertical spacing. As filler text, we will use a few sentences of the previous example about quoting, as follows:

1. Create a small document with the following code:

```
\documentclass{article}
\usepackage{parskip}
\usepackage{url}
\begin{document}
The authors of the CTAN team listed ten good reasons
for using \TeX. Among them are:

\TeX\ has the best output. What you end with,
the symbols on the page, is as useable, and beautiful,
as a non-professional can produce\ldots

The original text can be found on
```

```
\url{https://www.ctan.org/what_is_tex.html}.
\end{document}
```

2. Click **Typeset** and see the effect:

The authors of the CTAN team listed ten good reasons for using TEX. Among them are:

TEX has the best output. What you end with, the symbols on the page, is as useable, and beautiful, as a non-professional can produce...

The original text can be found on https://www.ctan.org/what_is_tex.html.

Figure 2.21 – Vertical spacing between paragraphs

Here, we loaded the `parskip` package to remove the paragraph indentation completely. Instead, this package adds vertical space between paragraphs. However, this package doesn't alter the behavior of the `quotation` environment; you can still use the `quote` environment.

There are two common ways to distinguish paragraphs. One is to indent the first line of each paragraph, which is LaTeX's default behavior. The other is to insert vertical space between paragraphs without indentation. This is often preferred for narrow columns, where indenting would occupy valuable space.

So far, we've practiced using many predefined **physical commands** to customize text and paragraphs. The next step is to create our own **logical formatting commands**, and use those within the document body text instead.

Creating commands

If you find yourself using the same term repeatedly in your document, it can get tedious to type it out each time. What if you later decide to change that term or its formatting? To avoid searching for and replacing the term in the entire document, LaTeX allows you to define your own custom commands in your preamble. That saves time and ensures consistency.

Remember: A command that's made up of other commands or text is called a **macro**, and that's what we'll define now. Basically, we choose a new macro name and define the sequence of text or commands to be used in that macro. Then, any time we want to reuse that content or formatting, we just need to call the macro by its name.

We'll begin with basic macros that work like abbreviations.

Using macros for simple text

Macros are great for avoiding repeating long words or phrases and can also act as placeholders, such as for person or company names or anything that can change in a different context. We can modify the macro's content to update the entire document with a different version of that phrase.

In the following example, we'll define a short command to represent the name of the **TeX Users Group** (**TUG**):

1. Enter the following code into a new document:

```
\documentclass{article}
\newcommand{\TUG}{\TeX\ Users Group}
\begin{document}
\section{The \TUG}
The \TUG\ is an organization for people who use
\TeX\ or \LaTeX.
\end{document}
```

2. Click **Typeset** and look at the result:

1 The TEX Users Group

The TEX Users Group is an organization for people who use TEX or LATEX.

Figure 2.22 – Using our first macro

We called `\newcommand` in the highlighted line to define our command. The first argument is the command's name we've chosen, and the second argument is the text it should expand to whenever we use that command in the document.

Now, every time we write `\TUG` in our document, LaTeX will print the complete name **TeX Users Group**. If we later decide to change the name or its formatting, we just need to change this `\newcommand` line. Then, the change of that macro will be applied to the entire document.

You may use formatting commands inside your command definition. Let's say you would like to change the formatting of all occurrences of this name to be typeset in small caps; just change the definition to the following:

```
\newcommand{\TUG}{\textsc{TeX Users Group}}
```

Notice that we've used the `\TeX` command. This abbreviation command just prints out the name of the typesetting system, formatted in the same way as its logo. The built-in `\LaTeX` command works similarly.

Note that we used a backslash after `\TeX`. The following space itself would just separate the command from the following text; it wouldn't produce a space in the output. Using the backslash followed by a space forces the output of a space that would otherwise be ignored. That also applies to the command we just created.

Now we will see how to avoid that manual spacing.

Improving spacing after commands

It's easy to forget the extra backslash needed to insert a space after a macro. Can we automate that? Tasks such as this, which aren't handled directly by LaTeX, can often be solved by using **packages**, which are collections of styles and commands.

In this case, we will use the `xspace` package; its only purpose is to adjust the spacing after a macro output automatically:

1. Insert this line into your **preamble**, before `\begin{document}`:

```
\usepackage{xspace}
```

2. Add the `\xspace` command to the end of your macro definition:

```
\newcommand{\TUG}{\TeX\ Users Group\xspace}
```

The `\usepackage{xspace}` line tells LaTeX to load the `xspace` package and to import its definitions. After `\usepackage`, we can use all commands contained in that package.

This package provides the `\xspace` command, which inserts a space depending on what the following character is:

- If a normal letter follows, then it will insert a space after the macro content
- If punctuation such as a dot, a comma, an exclamation mark, or a quotation mark follows, it won't insert a space

This simple automated solution works well in normal text; it always inserts a space if a letter follows and skips it when it detects punctuation. In a rare case that it's followed by something else, such as a macro that `xspace` cannot interpret, and inserts a space, you can prevent that by adding `{}` right after the `\xspace` command. The package manual explains it clearly; you can find it at `https://texdoc.org/pkg/xspace`.

Creating flexible commands with arguments

Suppose your document contains a lot of keywords that you want to be displayed in **bold**. If you use the `\textbf` command every time, what will happen if you later decide to use an *italic* shape instead, or a `typewriter` font? You'd have to update that formatting for every instance of that keyword, which is tedious and error-prone.

There's a better way: define a custom macro that uses `\textbf` internally.

Defining a macro with arguments

In this section, we'll use `\newcommand` again, but this time, we'll give it an argument that shall contain our keyword. For our example, we'll apply it to some terms that we learned about earlier in this chapter.

Let's get started:

1. Enter the following code in your editor. Our command will be called `\keyword`:

```
\documentclass{article}
\newcommand{\keyword}[1]{\textbf{#1}}
\begin{document}
\keyword{Grouping} by curly braces limits the
\keyword{scope} of \keyword{declarations}.
\end{document}
```

2. Click **Typeset** and observe how the keywords appear in the output:

Grouping by curly braces limits the **scope** of **declarations**.

Figure 2.23 – Formatting keywords

Let's look at the `\newcommand` line in the code. The number `1` in the square brackets marks the number of arguments that we want to use in the command. `#1` is a placeholder and will be replaced by the value of the first argument; `#2` would be replaced by the value of the second argument, and so on. Now, if you decide to change the appearance of all keywords to be *italics* instead, just modify the definition of `\keyword` to use `\textit{#1}`, and the change will be global.

The first time we used `\newcommand`, in the *Using macros for simple text* section, we used it with two arguments: the macro name and its definition. In the previous example, there were three arguments; the additional argument is enclosed in square brackets, which indicates that it is an **optional** argument, so that it may be provided or omitted. If omitted, it would have a default value.

Previously, we've already worked with the `\documentclass` command and optional arguments, but how can we *define* a command with optional arguments ourselves? Let's find out.

Defining a macro with optional arguments

Let's revisit `\newcommand`, but this time we'll give it an **optional** formatting argument along with a **mandatory** keyword argument:

1. Update the previous example to the following code:

```
\documentclass{article}
\newcommand{\keyword}[2][\bfseries]{{#1#2}}
\begin{document}
\keyword{Grouping} by curly braces limits the
\keyword{scope} of \keyword[\itshape]{declarations}.
\end{document}
```

2. Click on **Typeset** and check out the result:

Grouping by curly braces limits the **scope** of *declarations.*

Figure 2.24 – Optional arguments

Let's look again at the `\newcommand` line in the code. By using `[\bfseries]` in square brackets, we introduced an optional parameter; we refer to it by `#1`, and its default value is set to `\bfseries`. Since we used a **declaration** this time, we added a pair of braces to ensure that only the keyword is affected by the declaration. Later in the document, we passed `[\itshape]` to the `\keyword` command, changing the default formatting to italics.

Here's the general form of `\newcommand`:

```
\newcommand{command}[arguments][optional]{definition}
```

These are the meanings of the parameters to `\newcommand`:

- `command`: The new command's name, starting with a backslash followed by lowercase and/or uppercase letters, or a backslash followed by a single non-letter symbol. The name must not be already defined and is not allowed to begin with `\end`.
- `arguments`: A number from 1 to 9 that tells LaTeX how many arguments the new command accepts. If omitted, the command takes no arguments.
- `optional`: If present, the first of the arguments would be optional with a default value given here. Otherwise, all arguments are mandatory.

- `definition`: Every occurrence of `command` will be replaced by `definition` and every occurrence of the form `#n` will then be replaced by the nth argument.

Use `\newcommand` to define reusable styles for keywords, code snippets, web addresses, names, notes, information boxes, or differently emphasized text. That's how we achieved the consistent structure of this book, by defining and using styles. We should use font commands within our macro definitions rather than scattering them throughout the document body text.

Whenever possible, define your own macros to achieve a logical structure. You'll benefit from consistent formatting, easier global changes, and more maintainable code. By defining and using macros, you can ensure that the formatting remains consistent throughout your entire document.

Redefining commands

You can change existing LaTeX macros using `\renewcommand` in a very similar way:

```
\renewcommand{command}[arguments][optional]{definition}
```

You may rarely need it as a beginner, though it's a common way to change, for example, LaTeX's default values. Here are some examples you may encounter later:

- `\renewcommand{\familydefault}{\sfdefault}` makes the default font family the default sans-serif family, so that it is not a serif font by default anymore.
- `\renewcommand{\arraystretch}{1.5}` stretches the default row height in tables by a factor of 1.5.
- `\renewcommand{\thepage}{\Roman{page}}` switches the page number display to uppercase Roman numbers. `\roman` would mean lowercase.

This command is particularly useful if you want to customize an existing macro or command.

Creating environments

Just like `\newcommand` lets us define new commands, LaTeX has a command for defining blocks of content: `\newenvironment`. This is useful when we want to mark up larger sections of text, such as notes, boxes, examples, or other repeated layout patterns, with consistent formatting and structure.

Let's immediately jump to the general form:

```
\newenvironment{name}[arguments][optional]{begin-code}{end-code}
```

These are the meanings of the parameters to `\newenvironment`:

- `name`: The environment's name, used later as `\begin{name}` and `\end{name}`.
- `arguments`: An optional number from 1 to 9 indicating how many arguments the environment accepts; `0` by default when omitted.
- `optional`: If given, this assigns a default value to the first optional argument.
- `begin-code`: LaTeX will execute this code at the start of the environment. Use `#1`, `#2`, and so on to refer to the arguments.
- `end-code`: LaTeX runs this code when the environment ends.

For a very thorough specification of this and all macro definition commands, please refer to the reference at `https://latex2e.org/Definitions.html`.

Here's an example to illustrate it in practice:

```
\documentclass{article}
\newenvironment{note}[1][Note]
  {\begin{quote}\textbf{#1}\itshape}
  {\end{quote}}
\begin{document}
\begin{note}[Hint]
 Use high-contrast colors for readability.
\end{note}
\end{document}
```

This code does the following:

- It defines a `note` environment that takes an optional argument.
- It is based on the `quote` environment.
- If no argument is given, it prints **Note** at the beginning. Otherwise, it starts with the optional argument, such as **Hint** here.
- It continues with italic font for the remaining text in the environment.

Similar to `\renewcommand`, you can use `\renewenvironment` with exactly the same syntax to redefine an existing environment. That's something you'll rarely need, and you can also find it at `https://latex2e.org`, so we'll wrap it up at this.

Using an alternative syntax

LaTeX now provides additional ways to define commands and environments. You can continue to use the classic commands from the previous sections and skip this section for now. If you want, let's take a quick look at the newer approach. You can now define commands in this way:

```
\NewDocumentCommand{<command name>}{<argument types>}{code}
```

Argument types can be `m` for a **mandatory** argument, a lowercase `o` for an **optional** argument, or a capital `O{text}` for an optional argument with default text when the argument is omitted. But let's see how it looks in practice.

The definition for the macro in *Figure 2.22* would now be as follows:

```
\NewDocumentCommand{\TUG}{}{\TeX\ Users Group}
```

In the macro for *Figure 2.23*, we had one mandatory argument. It translates to the following:

```
\NewDocumentCommand{\keyword}{m}{\textbf{#1}}
```

For *Figure 2.24*, we added an optional argument with a default value. The same macro can also be written like this:

```
\NewDocumentCommand{\keyword}{O{\bfseries}m}{{#1#2}}
```

If you're curious about how the `o` option can be used to achieve the same, it looks as follows:

```
\NewDocumentCommand{\keyword}{om}{{%
  \IfNoValueTF{#1}%
    {\bfseries#2}%
    {#1#2}%
}}
```

Here, we used this command to handle both cases—when the optional argument is given and when it's not:

```
\IfNoValueTF{#1}%
  {Do stuff with argument #2 only}%
  {Do stuff with arguments #1 and #2}%
```

Note that I used the percent sign (%) at the end of lines to avoid accidentally introducing white space, since line breaks are treated as spaces. That's why I commented out the line breaks.

`TF` stands for **true and false**. There are also shorter variants for single cases. To just cover the case that no argument #1 was given (frue), the following is used:

```
\IfNoValueT{#1}{Do stuff with argument #2 only}%
```

If an argument #1 was given (false), the following is used:

```
\IfNoValueF{#1}{Do stuff with arguments #1 and #2}%
```

It might look a bit complex at first, but this system gives you much more flexibility. You can even define multiple optional arguments. For example, here's a more advanced definition with five mandatory and three optional ones:

```
\NewDocumentCommand{\name}{ m m m m m O{#3} O{#4} O{#5} }{ ... }
```

You see that you can have spaces between the specifiers for clarity, and you can use `#n` for a default value with a given nth argument. There are more argument types, and you can add delimiters for arguments, and you even have more similar commands, such as `\RenewDocumentCommand`, `\NewDocumentEnvironment`, and `\RenewDocumentEnvironment`, as replacements for the respective commands in the previous sections.

If you consider taking this new approach, have a look at these links:

- `https://www.texdev.net/2010/05/23/from-newcommand-to-newdocumentcommand/` – a blog article by LaTeX developer Joseph Wright with clear examples
- `https://texdoc.org/pkg/xparse` – the complete documentation
- `https://www.latex-project.org/publications/2021-JAW-TUB-tb130wright-newdoccmd.pdf` – a detailed comparison with other syntaxes and packages, for advanced users

Why might you want to use this approach?

- The LaTeX project team recommends it
- While it's easy to use for simple cases, you can define more complex macros
- Most importantly, in contrast to `\newcommand`, your new macros are **robust**, also referred to as **protected**; they aren't expanded to their definition immediately when LaTeX processes your code, but at a later stage

So, for example, if you use such a macro in a section title, the macro name would be written into the `.toc` (table of contents) file, while a `\newcommand` defined macro would be expanded immediately and its expanded content would go into the `.toc` file. Or, let's just say, robust commands often save you from subtle issues. It's a rather advanced topic; you may read about it at `https://texfaq.org/FAQ-protect`.

Personally, I'm fine with the original `\newcommand` and its companions, but I want you to know about the modern alternatives you will also see in newer LaTeX code on the internet.

Summary

In this chapter, we explored the fundamentals of editing, arranging, and formatting text. Specifically, we covered modifying fonts and styles, using commands and declarations with mandatory and optional arguments, and defining custom commands and environments. We also learned how to format paragraphs with different alignments, such as left, right, or fully justified, and how to include quotations.

Keep in mind that even though we used formatting commands directly in the text for practicing, it's better to use them in command definitions in the preamble to have styles consistently across a document. As we progress through this book, you'll discover further useful commands and packages to refine and extend your formatting setup.

Now that you've mastered detailed text formatting, you're ready to move on to the next chapter, which focuses on the formatting and layout of entire pages, including margins, headers, and footers.

3

Designing Pages

Now that you know how to format text, it's time to look at entire pages and the overall document structure.

In this chapter, we'll structure a document using chapters and sections, and learn how to adjust its page layout, including margins, page orientation, headers, and footers. This gives you control over the entire document design.

We'll cover the following topics:

- Structuring with chapters
- Adjusting page margins
- Creating a two-column layout
- Designing headers and footers
- Adding footnotes
- Handling page breaks
- Enlarging a page
- Changing line spacing
- Aligning the text area height
- Creating a table of contents

Working through these will build a better understanding of document classes and packages.

We'll start with a multi-page example document that we'll keep using throughout the chapters to try out new features. You can view, edit, and compile all examples online at the book's companion website: `https://latexguide.org/chapter-03`.

Structuring with chapters

Let's begin writing a longer document. We'll select the `book` document class and fill the pages with placeholder text to explore the page layout:

1. Create a new document, and enter the following lines as our preamble:

```
\documentclass[a4paper,12pt]{book}
\usepackage[english]{babel}
\usepackage{blindtext}
```

2. Proceed with writing the document body containing a chapter heading, section and subsection headings, and some filler text:

```
\begin{document}
\chapter{Exploring the page layout}
In this chapter we will study the layout of pages.
\section{Some filler text}
\blindtext
\section{A lot more filler text}
More dummy text will follow.
\subsection{Plenty of filler text}
\blindtext[10]
\end{document}
```

3. Compile it by clicking on **Typeset**. Look at the first page:

Chapter 1

Exploring the page layout

In this chapter we will study the layout of pages.

1.1 Some filler text

Hello, here is some text without a meaning. This text should show what a printed text will look like at this place. If you read this text, you will get no information. Really? Is there no information? Is there a difference between this text and some nonsense like "Huardest gefburn"? Kjift – not at all! A blind text like this gives you information about the selected font, how the letters are written and an impression of the look. This text should contain all letters of the alphabet and it should be written in of the original language. There is no need for special content, but the length of words should match the language.

1.2 A lot more filler text

More dummy text will follow.

1.2.1 Plenty of filler text

Hello, here is some text without a meaning. This text should show what a printed text will look like at this place. If you read this text, you will get no information. Really? Is there no information? Is there a difference between this text and some nonsense like "Huardest gefburn"? Kjift – not at all! A blind text like this gives you information about the selected font, how the letters are written and an impression of the look. This text should contain all letters of the alphabet and it should be written in of the original

1

Figure 3.1 – An example page

We have chosen the book document class. This class is suitable for extensive documents such as a book, a scientific report, or a thesis, when it's printed two-sided. Books are commonly two-sided and consist of chapters. By default, and this is very common, chapters start on right-hand pages, which have odd page numbers. If necessary, LaTeX inserts a blank, left-hand, even-numbered page so that the next chapter can begin on the following right-hand page.

Furthermore, books often have front matter, which includes one or more title pages, and back matter, which typically consists of a bibliography, index, and other relevant content. The book class supports all of this.

We used the `a4paper` option so the document would be formatted to fit on A4 paper. For the US letter paper size, we would use the `letterpaper` option instead. The document class `12pt` option instructed LaTeX to use a base font size of 12pt.

We loaded the `babel` package. That package provides typographic support tools for many languages, including proper hyphenation rules for the chosen language and translations for implicit terms. For example, we used the `english` option with `babel` and got **Chapter 1** in our chapter heading. If we choose `french` instead of `english`, we will get **Chapitre 1** in our heading.

American English is the default. For British English, we would use the `british` option with `babel`. There are very small differences between the two. In British English, some words are spelled differently, and hyphenation rules are a bit different, for example.

We loaded the `blindtext` package, which has been developed to produce filler text. It uses `babel` to detect the language of the document; we stated the `english` language to `babel`, which actually means American English. Without `babel`, `blindtext` would use Latin filler text by default. Then, the `\blindtext` command prints some dummy text just to fill the space.

The `\chapter` command produced a large heading, which will always begin on a new page. We've already seen the `\section` command. It's our second sectioning level and generates a smaller heading than `\chapter`. The numbering of this heading is automatically updated by LaTeX. Lastly, we refined the sectioning with the `\subsection` command followed by more dummy text to fill up the page.

Another widely used package for dummy text is called `lipsum`. It generates the famous *Lorem Ipsum* text that typesetters have used for decades.

Let's now look at how to adjust the default margins.

Adjusting page margins

Sometimes, publishers or supervisors request that you follow their specifications for a document. Apart from font size, interline spacing, and other style issues, there may be specifications for the margins. In this case, you would need to override LaTeX's defaults, specifying the margins precisely.

The `geometry` package makes that easy. We'll load the `geometry` package and state the exact width and height of all margins:

1. Extend the preamble of the previous example in this chapter with this command:

```
\usepackage[a4paper, inner=1.5cm, outer=3cm, top=2cm, bottom=3cm,
  bindingoffset=0.5cm]{geometry}
```

2. Click **Typeset** to compile the code and observe how the margins change.

The geometry package handles the layout, including paper size, margins, and other dimensions. We chose the A4 paper size, with an outer margin of 3 cm and an inner margin of 1.5 cm.

When a two-sided book lies open in front of us, the two inner margins visually merge into a single central space. To achieve visually balanced margins—left, center, and right—we can set the inner margin to half the width of the outer margin. That's why outer margins are typically wider than inner margins. There may be a reason to make the inner margin slightly wider; we might lose this space later due to binding methods such as gluing or stapling. But this depends on the kind of binding, and then it's done with an extra `bindingoffset` option.

We set the top margin to 2 cm and the bottom margin to 3 cm. Lastly, we specified a value of `0.5` cm for a binding correction.

In the early days of LaTeX, it was common to manipulate the layout dimensions directly. This approach had some disadvantages. We could easily make mistakes in calculating the lengths, for instance, the left margin plus the right margin plus the text width might not fit the paper width.

The `geometry` package simplifies this process. It provides an easy-to-use interface for layout settings. It provides auto-completion, calculates missing values to match the paper size, and even adds missing lengths using a heuristic approach to achieve a well-balanced layout.

The `geometry` package understands options as "`key=value`" pairs, separated by commas. If you load `geometry` without any arguments, you can set those arguments later with the `\geometry{argument list}` command.

Let's take a closer look at such `geometry` package options to control every aspect of your page layout.

Choosing the paper size

The `geometry` package offers various options for setting paper size and orientation:

- `paper=name` sets the paper size by name, for example, `paper=a4paper`. Supported sizes include `letterpaper`, `executivepaper`, `legalpaper`, `a0paper`, `a6paper`, `b0paper`, `b6paper`, and others.
- `paperwidth` and `paperheight` set custom paper dimensions, such as `paperwidth=7in` and `paperheight=10in`.

- `papersize={width,height}` sets both the width and height of the paper in one key, such as `papersize={7in,10in}`. This is an example of a double-valued argument.
- `portrait` sets the paper orientation to portrait mode, which is the default option. `landscape` switches the paper orientation to landscape mode.

If you have already specified the paper name in the document class, `geometry` will pick that up automatically. In fact, all document class options are passed to the packages that support them.

Specifying the text area

Use these options to control the size of the text block:

- `textwidth` sets the width of the main text area, such as `textwidth=140mm`
- `textheight` sets its height, such as `textheight=180mm`
- `lines` offer an alternative way to define the text height by specifying the number of lines, such as `lines=25`
- `includehead` includes the page header in the text height calculation; this option is set to `false` by default
- `includefoot` includes the footer of the page in the text area; this option is also set to `false` by default

Setting the margins

You can control the visible margins using the following options:

- `left` and `right` define the width of the left and right margins, such as `left=2cm`. Use these for one-sided documents.
- `inner` and `outer` set the width of the inner and the outer margin, such as `inner=2cm`. Use them for two-sided documents.
- `top` and `bottom` set the height of the vertical margins, such as `top=25mm`.
- `twoside` enables two-sided layout mode. This means that the left and right margins will be swapped on left-hand pages, also called verso pages.
- If your book is printed and bound, whether by glue, staples, or other means, the binding may cover part of the inner margin. You can set a value to the `bindingoffset` option to reserve width to compensate for the part of the inner margin that's hidden in the binding, so the visible inner margin looks as wide as you expect.

That's just a selection of commonly used options—there are many more. You can choose and set some options intuitively—for instance, `\usepackage[margin=3cm]{geometry}` will result in a 3 cm margin on each edge of the paper, and the paper size comes from the document class, unless specified otherwise.

The auto-completion works like this:

- `paperwidth = left + width + right`, where `width=textwidth` by default
- `paperheight = top + height + bottom`, where `height=textheight` by default

If you decide to include marginal notes within the text body when calculating the layout, the width could get wider than `textwidth`. If two dimensions of the right side of each formula are given, the missing dimension will be calculated. That's why it may be enough to specify `left` and `right`, and `top` and `bottom`, respectively. Even if just one margin is specified, the other dimensions will be determined using default margin ratios, as follows:

- `top:bottom = 2:3`
- `left:right = 1:1` for one-sided documents
- `inner:outer = 2:3` for two-sided documents

It might sound complex, but that's just how `geometry` helps you achieve a clean, balanced page layout, even when some values are missing.

The `geometry` package comes with a comprehensive manual. Don't let its length intimidate you; it's designed to guide you through the many features.

As shown earlier in *Chapter 1, Getting Started with LaTeX*, you can open the manual by typing `texdoc geometry` at the command line, a terminal window, or on the internet at `https://texdoc.org/pkg/geometry`.

Now that you know how to set up the basic page geometry, we'll move on to options for changing the text layout, such as landscape orientation and multi-column formatting.

Creating a two-column layout

You already know that a document class defines the foundation of your document. It provides commands and environments extending the LaTeX standard features. Though the class provides a default style, you can customize it using **document class options**.

Let's switch the orientation of our first example to landscape, and typeset our text in two columns:

1. Add the `landscape` and `twocolumn` options to the `\documentclass` line of our example, as follows:

```
\documentclass[a4paper,12pt,landscape,twocolumn]{book}
```

2. Load the `geometry` package:

```
\usepackage{geometry}
```

3. Click **Typeset** to compile, and see the new layout in effect:

Chapter 1

Exploring the page layout

In this chapter we will study the layout of pages.

1.1 Some filler text

Hello, here is some text without a meaning. This text should show what a printed text will look like at this place. If you read this text, you will get no information. Really? Is there no information? Is there a difference between this text and some nonsense like "Huardest gefburn"? Kjift – not at all! A blind text like this gives you information about the selected font, how the letters are written and an impression of the look. This text should contain all letters of the alphabet and it should be written in of the original language. There is no need for special content, but the length of words should match the language.

1.2 A lot more filler text

More dummy text will follow.

1.2.1 Plenty of filler text

Hello, here is some text without a meaning. This text should show what a printed text will look like at this place. If you read this text, you will get no information. Really? Is there no information? Is there a difference between this text and some nonsense like "Huardest gefburn"? Kjift – not at all! A blind text like this gives you information about the selected font, how the letters are written and an impression of the look. This text should contain all letters of the alphabet and it should be written in of the original language. There is no need for special content, but the length of words should match the language. Hello, here is some text without a meaning.

Figure 3.2 – A landscape two-column page layout

By using the `landscape` option, we switched the page orientation from portrait to landscape. The `twocolumn` option splits the document body text into two columns.

We loaded the `geometry` package to ensure that the PDF page matches the landscape layout. Without it, the final PDF would remain in portrait mode.

The `\twocolumn[opening text]` command starts a two-column page with optional opening text over the full width. `\onecolumn` begins a one-column page. If you'd like to balance the columns on the last page or if you wish to have more than two columns, use the `multicols` package.

The LaTeX base classes are `article`, `book`, `report`, `slides`, and `letter`. As the name suggests, the last one can be used to write letters, though there are further suitable classes, such as `scrlttr2`.

`slides` can be used to create presentations, but today there are more powerful and feature-rich classes, such as `beamer` and `powerdot`.

Let's sum up the options of the base classes:

- `a4paper`, `a5paper`, `b5paper`, `letterpaper`, `legalpaper`, or `executivepaper`: The output will be formatted according to this paper size; for example, A4 will be formatted as 210 mm x 297 mm. The `letterpaper` option (8.5 in x 11 in) is the default. Loading the `geometry` package allows for more sizes.
- `10pt`, `11pt`, or `12pt`: The size of normal text in the document; the default is 10 points (`10pt`). The size of headings, footnotes, indexes, and so on, will be scaled accordingly.
- `landscape`: Switches to landscape format by swapping the width and height of the output paper dimensions.
- `onecolumn` or `twocolumn`: Decides whether the pages will be one-column (default) or two-column. The `letter` class doesn't support it.
- `oneside` or `twoside`: Controls layout for printing on one side or both sides of a page. `oneside` is the default, except for the `book` class. `twoside` is not applicable to the `slides` class and the `letter` class.
- `openright` or `openany`: The first option lets chapters begin on a right-hand page (the default for the `book` class), while the second option permits chapters to start on any page (the default for the `report` class). These options are only supported by the `book` and `report` classes because the other classes don't provide chapters.
- `titlepage` or `notitlepage`: The first causes a separate title page when `\maketitle` is used and is the default, except for the `article` class. The default of `article` is `notitlepage`, which means that normal text may follow the title on the same page.
- `final` or `draft`: If `draft` is set, then LaTeX will mark overfull lines with a black box, which is helpful for reviewing and improving the output. Some packages support these options as well, behaving differently then, such as omitting the embedding of graphics and listings when a draft has been chosen. `final` is the default.
- `openbib`: When this option is set, a bibliography will be formatted in open style instead of compressed style.
- `fleqn`: This left-aligns displayed formulas instead of centering.
- `leqno`: For numbered displayed formulas, the number will be put on the left side. The right side is the default.

Many other classes support these options as well, and often offer even more. For an uncommon base font size, the `extarticle`, `extbook`, `extreport`, and `extletter` classes provide base font sizes from 8 points to 20 points. The **KOMA-Script** classes support arbitrary base font sizes.

KOMA-Script classes serve as enhanced alternatives to the base classes: for each base class, such as `article`, `book`, or `report`, there's a corresponding KOMA class (`scrartcl`, `scrbook`, and `scrreprt`, respectively). They offer a significantly expanded feature set, with many additional commands and options for customizing document layout and behavior. Visit `https://texdoc.org/pkg/koma-script` to consult the manual.

As page headers have been mentioned, let's explore them now.

Designing headers and footers

When we tested the initial version of our example, you may have noticed that, except on the chapter's first page, all other pages displayed the page number, chapter title, and section title in their headers. So, in our two-sided layout, on page **2**, in a left-hand page header, the page number is in the outer margin—here, on the left side:

2 *CHAPTER 1. EXPLORING THE PAGE LAYOUT*

language. There is no need for special content, but the length of words should match the language. Hello, here is some text without a meaning. This text should show what a printed text will look like at this place. If you read this text, you will get no information. Really? Is there no information? Is there a difference between this text and some nonsense like "Huardest gefburn"?

Figure 3.3 – The header of page 2

By contrast, this is how our right-hand page header on page **3** looks, with the page number in the outer margin, which is on the right side now:

1.2. A LOT MORE FILLER TEXT 3

language. There is no need for special content, but the length of words should match the language. Hello, here is some text without a meaning. This text should show what a printed text will look like at this place. If you read this text, you will get no information. Really? Is there no information? Is there a difference between this text and some nonsense like "Huardest gefburn"?

Figure 3.4 – The header of page 3

In a one-sided layout, there would be no such difference in the header layout; it would follow a uniform style. The headers in a one-sided layout are as in *Figure 3.4*. By default, heading text is on the left side, and the page number is on the right side.

Although these standard headers are already quite suitable, we can further customize them to suit our preferences better.

The default shape of the page headings is slanted. Furthermore, they are written in capital letters. We shall use bold typeface instead, and we will use a small-caps font for the chapter title. We will load the `fancyhdr` package and use its commands to achieve that:

1. Load the first example of this chapter.
2. Insert the highlighted lines to get this:

```
\documentclass[a4paper,12pt]{book}
\usepackage[english]{babel}
\usepackage{blindtext}
\usepackage{fancyhdr}
\fancyhf{}
\fancyhead[LE]{\scshape\nouppercase{\leftmark}}
\fancyhead[RO]{\nouppercase{\rightmark}}
\fancyfoot[LE,RO]{\thepage}
\pagestyle{fancy}
\begin{document}
\chapter{Exploring the page layout}
In this chapter we will study the layout of pages.
\section{Some filler text}
\blindtext
\section{A lot more filler text}
More dummy text will follow.
\subsection{Plenty of filler text}
\blindtext[10]
\end{document}
```

3. Compile the code. The footers will contain the page number on their outer side.

A right-hand page header now looks like the following:

CHAPTER 1. EXPLORING THE PAGE LAYOUT

language. There is no need for special content, but the length of words should match the language. Hello, here is some text without a meaning. This text should show what a printed text will look like at this place. If you read this text, you will get no information. Really? Is there no information? Is there a difference between this text and some nonsense like "Huardest gefburn"?

Figure 3.5 – The new header of page 2

A left-hand header now looks like this:

1.2. A lot more filler text

language. There is no need for special content, but the length of words should match the language. Hello, here is some text without a meaning. This text should show what a printed text will look like at this place. If you read this text, you will get no information. Really? Is there no information? Is there a difference between this text and some nonsense like "Huardest gefburn"?

Figure 3.6 – The new header of page 3

We used the `fancyhdr` package to customize the headers and footers of our document. The package command names start with `\fancy`. Our first action was calling `\fancyhf{}`; this command clears the headers and footers. Furthermore, we used the following commands:

- `\leftmark`: Stores the chapter title together with the chapter number. Capital letters are used as the default.
- `\rightmark`: Stores the section title together with its number. Capital letters are used as well.
- `\nouppercase`: This command prevents the automatic capitalization in its argument.
- `\scshape`: We switched to a small-caps font.

We used the `\fancyhead` command with the optional `LE` argument to place the chapter title in the header. `LE` stands for *left-even* and means that this chapter title will be put on the left side of the header on even-numbered pages.

Conversely, we called the `\fancyhead` command with `RO` to put the section title into the header. `RO` stands for *right-odd* and means that this section heading shall be displayed on the right side of the header on odd-numbered pages.

Afterward, we used `\fancyfoot` to display the page number in the footer. This time, we used `LE` and `RO`, which showed the page number on both even and odd pages, always on the outer side. Then, the `\thepage` command prints the page number.

All these commands are used to modify a page style provided by `fancyhdr`; this style is called `fancy`. We had to instruct LaTeX to use this style, and we did so through the `\pagestyle{fancy}` command.

Emphasizing by writing all letters capitalized, as `fancyhdr` does by default, is called **all caps**. It is widely regarded as a questionable style. That's why we moved to small caps.

There are different styles of headers and footers. That combination is referred to as a **page style**. Let's see what page styles are available.

Understanding page styles

LaTeX and its base classes offer four built-in page styles:

- `empty`: Neither a header nor a footer is shown.
- `plain`: No header. The page number will be printed and centered in the footer.
- `headings`: The header contains titles of chapters, sections, and/or subsections, depending on the class, as well as the page number. The footer remains empty.
- `myheadings`: The header contains user-defined text and the page number; the footer is empty.

The `fancyhdr` package introduces a powerful page style called `fancy`, which enables users to customize both the header and the footer.

The following two commands may be used to choose the page style:

- `\pagestyle{name}`: Switches to the page style name from this point onward
- `\thispagestyle{name}`: Applies the `name` page style only to the current page; subsequent pages will retain the style used previously

You have noticed that the page style changes where a chapter begins, differing from the style of other pages. Such pages will have a `plain` style. If you think all pages should use the same style, consider some books: it's very common for chapter beginnings to differ in style. They usually have a blank header. The `\thispagestyle` command can be used to override that.

We can modify content and positioning in headers and footers, as we will see next.

Customizing headers and footers

LaTeX divides both headers and footers into three pieces: left, center, and right (`l`, `c`, and `r`, respectively). You can customize each part using the following commands:

- For the header: `\lhead`, `\chead`, or `\rhead`
- For the footer: `\lfoot`, `\cfoot`, or `\rfoot`

Each of these commands requires a mandatory argument, such as `\chead{User's guide}` or `\cfoot{ \thepage}`. This argument will be put into the corresponding area of the page.

Alternatively, you could use these versatile commands:

- Header: `\fancyhead[code]{text}`
- Footer: `\fancyfoot[code]{text}`

Here, code may consist of one or more letters:

- `L`: Left
- `C`: Center
- `R`: Right
- `E`: Even page
- `O`: Odd page

It doesn't matter if we choose uppercase or lowercase letters. We already used such combinations in our example.

Another customization is modifying the separation line between the text and the footer, up next.

Using decorative lines in headers or footers

We can add and adjust lines between the header and the body text, and the body text and the footer, respectively, with these two commands:

- `\renewcommand{\headrulewidth}{width}`
- `\renewcommand{\footrulewidth}{width}`

Here, `width` may be a value such as `1pt`, `0.5mm`, and so on. The default is `0.4pt` for the header line and `0pt` for the footer line. `0pt` means that a line is not visible.

While `\newcommand` defines a new command, `\renewcommand` redefines an existing command. Incidentally, we've got to know a new concept: a lot of LaTeX commands may be redefined in this way. This can be simply changing a value, as here, or redefining the code for a command.

Changing LaTeX's header marks

As we already know, LaTeX classes and packages automatically store sectioning numbers and headings in the `\leftmark` and `\rightmark` macros. It will be done when we call `\chapter`, `\section`, or `\subsection`. So, we could just use `\leftmark` and `\rightmark` in the arguments of the `fancyhdr` commands.

We may sometimes need to manually update those entries, even if we rely on this automation. For instance, the starred sectioning commands such as `\chapter*` and `\section*` won't produce a header entry, as indicated earlier. In such a case, two commands will help us:

- `\markright{right head}` sets the right heading
- `\markboth{left head}{right head}` sets both the left and right headings

The default `headings` style is easy to use and gives good results. `myheadings` can be used together with `\markright` and `\markboth`. However, the most flexible way is given by `fancyhdr`, especially in combination with `\markright` and `\markboth`.

A very good alternative to `fancyhdr` is the package called `scrpage-scrlayer`. It belongs to KOMA-Script, but `scrpage-scrlayer` works with other classes as well. It provides similar functionality and offers even more features.

The footer is a good place to add notes. Let's see how to do that in the next section.

Adding footnotes

As briefly mentioned in *Chapter 2, Formatting Text and Creating Macros,* LaTeX provides a command to typeset footnotes. Let's see it in action.

Let's go back to the very first example of this chapter. We shall insert one footnote in the body text and one in a section heading:

1. Modify the example by inserting a footnote, as shown in the highlighted line:

```
\documentclass[a4paper,12pt]{book}
\usepackage[english]{babel}
\usepackage{blindtext}
\begin{document}
\chapter{Exploring the page layout}
In this chapter we will study the layout of pages.
\section{Some filler text}
\blindtext
```

```
\section{A lot more filler text}
More dummy text\footnote{serving as a placeholder}
will follow.
\subsection{Plenty of filler text}
\blindtext[10]
\end{document}
```

2. Compile the code to see how the footnote looks in print:

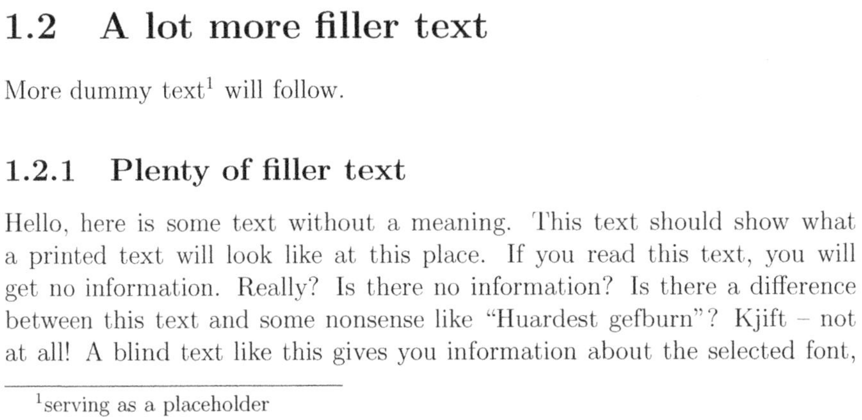

1.2 A lot more filler text

More dummy text[1] will follow.

1.2.1 Plenty of filler text

Hello, here is some text without a meaning. This text should show what a printed text will look like at this place. If you read this text, you will get no information. Really? Is there no information? Is there a difference between this text and some nonsense like "Huardest gefburn"? Kjift – not at all! A blind text like this gives you information about the selected font,

[1]serving as a placeholder

Figure 3.7 – Text with a footnote

The `\footnote{text}` command placed a superscripted number at the current position.

Furthermore, it prints its argument text at the bottom of the page, marked by the same number. As we've seen, such notes are separated from the main text by a horizontal line.

`\footnote[number]{text}` produces a footnote marked by this optional number, an integer. If we don't give the optional number, an internal counter will be stepped and used. This will be done automatically; we don't need to worry.

Two additional commands help us selectively put only a footnote mark or text:

- `\footnotemark[number]` produces a superscripted number in the text as a footnote mark. If the optional argument is not given, it also steps and uses the internal footnote counter. No footnote text will be generated.
- `\footnotetext[number]{text}` generates footnote text without putting a footnote mark in the text, and it does not step up the internal footnote counter.

Set a `footnote` command right after the related text. Don't leave a space in between; otherwise, you would get a gap between the text and the following footnote mark.

In *Figure 3.7*, we saw a line that separates the footnotes from the text. We will now see how to adjust that line.

Modifying the footnote line

The line that separates footnotes from the text is produced by the `\footnoterule` command. If we wish to omit that line or modify it, we must redefine it. We learned about `\renewcommand` earlier, so let's use it.

We will use `\renewcommand` to override the default `\footnoterule` command:

1. Take the previous example and add the following lines to the preamble:

```
\renewcommand{\footnoterule}
{\noindent\smash{\rule[3pt]{\textwidth}{0.4pt}}}
```

2. Click on **Typeset** to compile, and see how the line has changed:

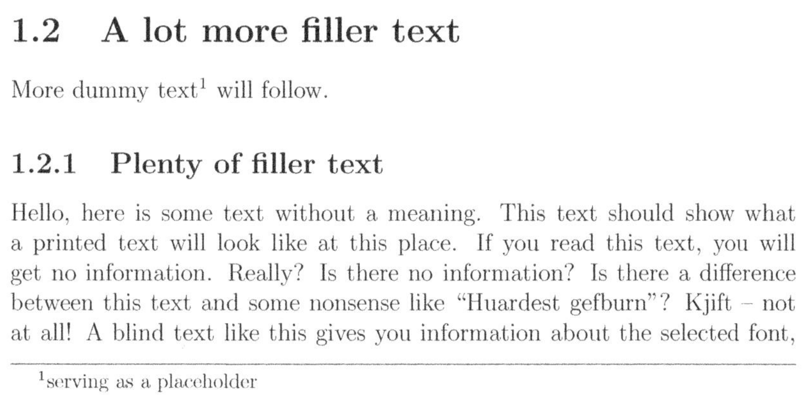

1.2 A lot more filler text

More dummy text[1] will follow.

1.2.1 Plenty of filler text

Hello, here is some text without a meaning. This text should show what a printed text will look like at this place. If you read this text, you will get no information. Really? Is there no information? Is there a difference between this text and some nonsense like "Huardest gefburn"? Kjift – not at all! A blind text like this gives you information about the selected font,

[1]serving as a placeholder

Figure 3.8 – A modified footnote line

The existing `\footnoterule` command will be replaced by the new definition that we wrote in the second line of the first step. The `\rule[raising]{width}{height}` command draws a line, here 0.4 pt thick, and as wide as the text, raised a bit, by 3 pt. Through the `\smash` command, we let our line pretend to have a height and depth of zero, so it occupies no vertical space at all. This way, the page balancing will not be affected. You already know `\noindent`, which avoids the paragraph indentation.

If you want to omit that footnote line completely, you just need to write the following:

```
\renewcommand{\footnoterule}{}
```

Now, the command is defined to do nothing, and we won't get a dividing line.

Using packages to expand footnote styles

There are different habits for setting footnotes. Some styles require footnotes numbered per page; they might have to be placed in the document as so-called endnotes, and symbols instead of numbers may be used. More demands exist, and, therefore, several packages have been developed to comply with them. Here's a selection:

- `endnotes`: Places footnotes at the end of the document
- `manyfoot`: Allows nested footnotes
- `bigfoot`: Replaces and extends `manyfoot` and improves page break handling with footnotes
- `savefnmark`: Useful when you need to use footnotes several times
- `footmisc`: An all-round package; introduces numbering per page, is able to save space when many short footnotes are used, offers symbols instead of numbers as footnote marks, and provides hanging indentation and other styles

Have a look at the respective package documentation to learn more using either the `texdoc` command, as explained in *Chapter 1, Getting Started with LaTeX*, or at `https://texdoc.org`.

As we've talked about footnotes at the end of a page, let's see how to enforce ending a page ourselves in case we don't want to let it automatically happen.

Handling page breaks

As you've seen in our example, LaTeX itself handles the page breaking. There might be occasions where we'd like to insert a page break ourselves before LaTeX does. LaTeX offers several commands to achieve this, with or without vertical balance.

We will now go back to the first version of our example, and we shall manually insert a page break right before subsection 1.2.1:

1. Insert the highlighted line into our example, which contains the `\pagebreak` command:

   ```
   \documentclass[a4paper,12pt]{book}
   \usepackage[english]{babel}
   \usepackage{blindtext}
   \begin{document}
   ```

```
\chapter{Exploring the page layout}
In this chapter we will study the layout of pages.
\section{Some filler text}
\blindtext
\section{A lot more filler text}
More dummy text will follow.
\pagebreak
\subsection{Plenty of filler text}
\blindtext[10]
\end{document}
```

2. Compile the code and take a look at the result:

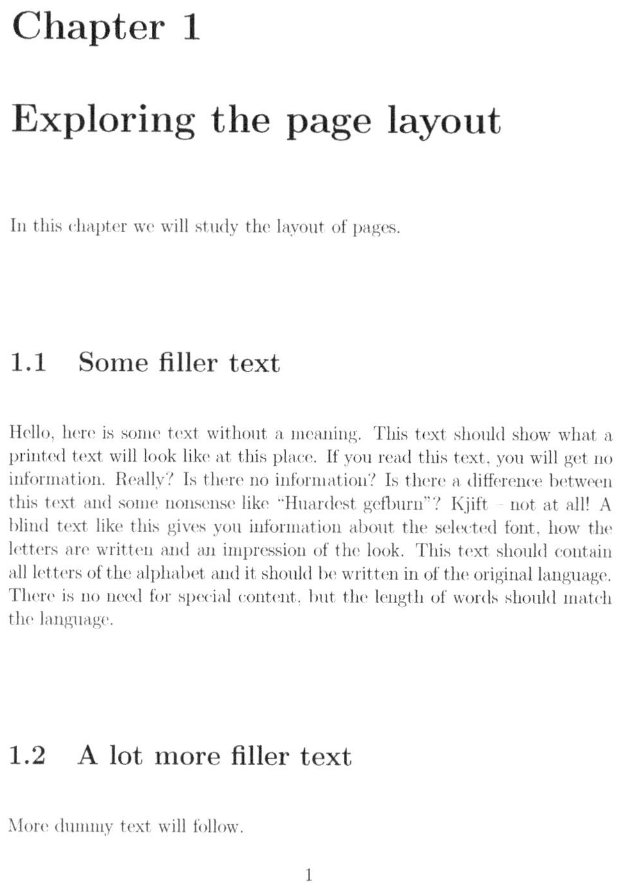

Chapter 1

Exploring the page layout

In this chapter we will study the layout of pages.

1.1 Some filler text

Hello, here is some text without a meaning. This text should show what a printed text will look like at this place. If you read this text, you will get no information. Really? Is there no information? Is there a difference between this text and some nonsense like "Huardest gefburn"? Kjift – not at all! A blind text like this gives you information about the selected font, how the letters are written and an impression of the look. This text should contain all letters of the alphabet and it should be written in of the original language. There is no need for special content, but the length of words should match the language.

1.2 A lot more filler text

More dummy text will follow.

1

Figure 3.9 – A stretched page

3. Replace `\pagebreak` with `\newpage`.

4. Compile again, and compare:

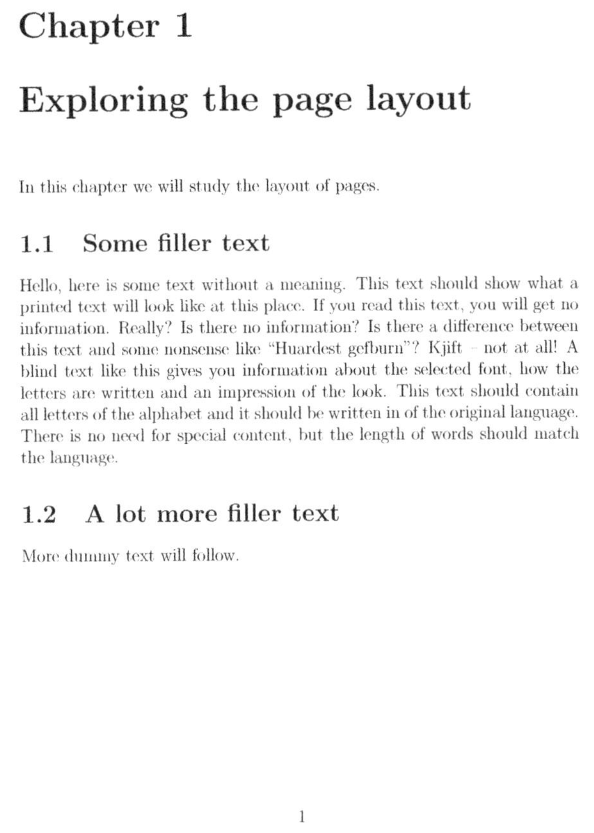

Chapter 1

Exploring the page layout

In this chapter we will study the layout of pages.

1.1 Some filler text

Hello, here is some text without a meaning. This text should show what a printed text will look like at this place. If you read this text, you will get no information. Really? Is there no information? Is there a difference between this text and some nonsense like "Huardest gefburn"? Kjift – not at all! A blind text like this gives you information about the selected font, how the letters are written and an impression of the look. This text should contain all letters of the alphabet and it should be written in of the original language. There is no need for special content, but the length of words should match the language.

1.2 A lot more filler text

More dummy text will follow.

1

Figure 3.10 – A non-stretched page

At first, we inserted the `\pagebreak` command; as its name suggests, it causes a page break. Furthermore, the text has been stretched to fill the page down to the bottom. That can be desirable for having the same text height on all pages.

Afterward, due to the obviously unpleasant whitespace between the paragraphs and headings, we replaced `\pagebreak` with `\newpage`. This command breaks the page as well, but it doesn't stretch the text: the remaining space of the page will stay empty.

So, `\pagebreak` behaves like `\linebreak`, and `\newpage` works like `\newline` (for pages instead of lines). There's even a `\nopagebreak` command that's analogous to `\nolinebreak` and forbids page breaking. `\pagebreak` won't break a line, while `\nopagebreak` doesn't refer to the middle of a line; both commands apply at the end of the current line. Of course, they immediately have an effect when used between paragraphs.

If you use the two-column format, both `\pagebreak` and `\newpage` will begin on a new column instead of a new page.

There are two further variants: `\clearpage` works like `\newpage`, except that it will start on a new page, even in two-column mode. `\cleardoublepage` does the same but causes the following text to start on a right-hand page, inserting a blank page if necessary. The latter is helpful for two-sided documents.

More importantly, both commands cause all figures and tables that LaTeX has in its memory to be printed out immediately.

`\pagebreak` and `\nopagebreak` can take an optional argument that requests a certain line break, as follows. The argument is an integer between `0` and `4`. Here, `0` means that a page break is allowed, `1` means it's desired, `2` and `3` mark more insistent requests so LaTeX tries harder to stretch the text to reach the page bottom, and `4` will enforce a page break. `\pagebreak` and `\nopagebreak` are very similar to the `\linebreak` and `\nolinebreak` command pair, which we saw in *Chapter 2, Formatting Text and Creating Macros*.

Such manual page breaks reduce the amount of text that fits on the page. Let's now examine the opposite: adding more text to a page.

Enlarging a page

There may be occasions where we want to add a little more text to a page, even if it will be squeezed a bit or the text height increases. There's a command that will help us out: `\enlargethispage`.

We'll modify our example slightly. This time, we will try to avoid a nearly empty page by squeezing the text on the preceding page:

1. Remove the `\newpage` command from our example and switch to an `11pt` base font. This time, use less filler text in the subsection:

   ```
   \documentclass[a4paper,11pt]{book}
   \usepackage[english]{babel}
   \usepackage{blindtext}
   \usepackage[a4paper, inner=1.5cm, outer=3cm, top=2cm,
   bottom=3cm, bindingoffset=1cm]{geometry}
   \begin{document}
   \chapter{Exploring the page layout}
   In this chapter we will study the layout of pages.
   \section{Some filler text}
   ```

```
\blindtext
\section{A lot more filler text}
More dummy text will follow.
\subsection{Plenty of filler text}
\blindtext[3]
\end{document}
```

2. Compile, and the result will consist of two pages. This is the first page:

Chapter 1

Exploring the page layout

In this chapter we will study the layout of pages.

1.1 Some filler text

Hello, here is some text without a meaning. This text should show what a printed text will look like at this place. If you read this text, you will get no information. Really? Is there no information? Is there a difference between this text and some nonsense like "Huardest gefburn"? Kjift – not at all! A blind text like this gives you information about the selected font, how the letters are written and an impression of the look. This text should contain all letters of the alphabet and it should be written in of the original language. There is no need for special content, but the length of words should match the language.

1.2 A lot more filler text

More dummy text will follow.

1.2.1 Plenty of filler text

Hello, here is some text without a meaning. This text should show what a printed text will look like at this place. If you read this text, you will get no information. Really? Is there no information? Is there a difference between this text and some nonsense like "Huardest gefburn"? Kjift – not at all! A blind text like this gives you information about the selected font, how the letters are written and an impression of the look. This text should contain all letters of the alphabet and it should be written in of the original language. There is no need for special content, but the length of words should match the language. Hello, here is some text without a meaning. This text should show what a printed text will look like at this place. If you read this text, you will get no information. Really? Is there no information? Is there a difference between this text and some nonsense like "Huardest gefburn"? Kjift – not at all! A blind text like this gives you information about the selected font, how the letters are written and an impression of the look. This text should contain all letters of the alphabet and it should be written in of the original language. There is no need for special content, but the length of words should match the language. Hello, here is some text without a meaning. This text should show what a printed text will look like at this place. If you read this text, you will get no information. Really? Is there no information? Is there a difference between this text and some nonsense like "Huardest gefburn"? Kjift – not at all! A blind text like this gives you information about the selected font, how the letters are written and an impression of the look. This text should contain all letters of the alphabet and it should be written in

1

Figure 3.11 – A fully filled page

And this is the text on the second page:

2 CHAPTER 1. EXPLORING THE PAGE LAYOUT

of the original language. There is no need for special content, but the length of words should match the language.

Figure 3.12 – Remaining text on the second page

3. Insert this command right after the `\subsection` line:

```
\enlargethispage{\baselineskip}
```

4. Compile again, and now our document fits on just one page:

Chapter 1

Exploring the page layout

In this chapter we will study the layout of pages.

1.1 Some filler text

Hello, here is some text without a meaning. This text should show what a printed text will look like at this place. If you read this text, you will get no information. Really? Is there no information? Is there a difference between this text and some nonsense like "Huardest gefburn"? Kjift – not at all! A blind text like this gives you information about the selected font, how the letters are written and an impression of the look. This text should contain all letters of the alphabet and it should be written in of the original language. There is no need for special content, but the length of words should match the language.

1.2 A lot more filler text

More dummy text will follow.

1.2.1 Plenty of filler text

Hello, here is some text without a meaning. This text should show what a printed text will look like at this place. If you read this text, you will get no information. Really? Is there no information? Is there a difference between this text and some nonsense like "Huardest gefburn"? Kjift – not at all! A blind text like this gives you information about the selected font, how the letters are written and an impression of the look. This text should contain all letters of the alphabet and it should be written in of the original language. There is no need for special content, but the length of words should match the language. Hello, here is some text without a meaning. This text should show what a printed text will look like at this place. If you read this text, you will get no information. Really? Is there no information? Is there a difference between this text and some nonsense like "Huardest gefburn"? Kjift – not at all! A blind text like this gives you information about the selected font, how the letters are written and an impression of the look. This text should contain all letters of the alphabet and it should be written in of the original language. There is no need for special content, but the length of words should match the language. Hello, here is some text without a meaning. This text should show what a printed text will look like at this place. If you read this text, you will get no information. Really? Is there no information? Is there a difference between this text and some nonsense like "Huardest gefburn"? Kjift – not at all! A blind text like this gives you information about the selected font, how the letters are written and an impression of the look. This text should contain all letters of the alphabet and it should be written in of the original language. There is no need for special content, but the length of words should match the language.

1

Figure 3.13 – All text fits on a single page

We used the `\enlargethispage` command to squeeze more text onto a page. This command takes the additionally requested height as its argument. The `\baselineskip` command returns the height of a text line that we used as the argument. So, LaTeX could put one extra line onto the page, and even the remaining line fitted in as well because LaTeX compressed some whitespace.

We could use factors: write `\enlargethispage{2\baselineskip}` to get two more lines on a page. It doesn't even need to be an integer value. As always, when stating a length, you can use other units, such as `10pt`, `0.5in`, `1cm`, or `5mm`, and even negative values.

Only the current page will be affected by this command. There's a starred version: `\enlargethispage*` would additionally shrink all vertical spaces on the page to their minimum.

However, `\enlargethispage` should be considered just for a possible easy fix when you quickly need to put more text on a single page. In general, we can adjust the text quantity on the page by changing the margins, as we already know, or by adjusting the line spacing within text. Let's examine line spacing in the next section.

Changing the line spacing

Without some vertical space between the lines, the readability of our text could suffer. Adding such space would help lead the eye along the line. Although LaTeX already ensures good readability by choosing meaningful interline spacing, publishers may require different spacing.

We shall modify the very first example of this chapter by adding half of a line height to the line spacing:

1. Extend the preamble of our example with this command:

```
\usepackage[onehalfspacing]{setspace}
```

2. Compile the code to see the change:

Chapter 1

Exploring the page layout

In this chapter we will study the layout of pages.

1.1 Some filler text

Hello, here is some text without a meaning. This text should show what a printed text will look like at this place. If you read this text, you will get no information. Really? Is there no information? Is there a difference between this text and some nonsense like "Huardest gefburn"? Kjift – not at all! A blind text like this gives you information about the selected font, how the letters are written and an impression of the look. This text should contain all letters of the alphabet and it should be written in of the original language. There is no need for special content, but the length of words should match the language.

1.2 A lot more filler text

More dummy text will follow.

1.2.1 Plenty of filler text

Hello, here is some text without a meaning. This text should show what a printed text will look like at this place. If you read this text, you will get no information. Really? Is there no information? Is there a difference between this text and some nonsense like "Huardest gefburn"? Kjift – not at all! A blind text like this gives you information about the selected font, how the letters are written and an impression of the look. This text should contain all letters of the alphabet and it should be written in of the original language. There is no need for special content, but the length of words should match the language. Hello, here is some text without a meaning. This text should show what a printed text will look like at this place. If you read this text, you will get no information. Really? Is there no information? Is there a difference between this text and some nonsense like "Huardest gefburn"? Kjift – not at all! A blind text like this gives you information about the selected font, how the letters are written and an impression of the look. This text should contain all letters of the alphabet and it should be written in of the original language. There is no need for special content, but the length of words should match the language. Hello, here is some text without a meaning.

1

Figure 3.14 – Additional interline spacing

We loaded the `setspace` package to adjust the line spacing. We provided the `onehalfspacing` option, which increases the spacing by half of a line height for the whole document.

The `setspace` package understands three options:

- `singlespacing` is the default. No additional space will be inserted. The text will be typeset with LaTeX's default interline spacing, which is about 20 percent of the line height.
- `onehalfspacing` means one-and-a-half spacing, as you can see in our example.
- `doublespacing` can be used for even more spacing; the distance between the baselines of successive text lines would be twice as high as a single line.

In typesetter's jargon, the distance between the baselines of consecutive text lines is referred to as **leading** (pronounced "ledding" from the metal "lead," once used to separate lines).

Now that we adjusted vertical spacing and page breaks, let's see how to align the text area on the pages vertically.

Aligning the text area height

Let's revisit *Figure 3.9*. It showed a page where the text appeared stretched: extra vertical space had been inserted between paragraphs and headings. This behavior is typical of the `book` document class: in two-sided layouts, LaTeX aligns the text block vertically so that facing pages have the same height.

While this balancing is intended to be visually pleasing, it may sometimes lead to awkward spacing, especially if the page contains little text. In such cases, you might prefer to disable this alignment.

LaTeX offers two commands to control vertical text alignment:

- `\raggedbottom` disables vertical alignment, so you get less stretching and natural page height
- `\flushbottom` enforces equal text height on every page, so LaTeX stretches the content if necessary

These commands are usually placed in the preamble for a consistent layout across the document. However, they can also be used selectively within the document to switch styles when needed.

Here's a practical guideline:

- Use `\flushbottom` for text-rich, two-sided documents such as books for a perfect visual appearance when the book is opened. That's the default with the `book` class.

- Use `\raggedbottom` when your content varies in density, such as when you have many figures and tables and notice unwanted spacing. That's the default in one-sided document classes such as `article`, `report`, `letter`, and `presentation` classes.

Now that we have finished designing the entire document, let's add a table of contents.

Creating a table of contents

A book commonly begins with a table of contents, so let's create one based on our numbered headings:

1. In our previous document, let's remove the `landscape` and `twocolumn` options.
2. Remove the `setspace` package, that is, delete this line:

```
\usepackage[onehalfspacing]{setspace}
```

3. Add the `\tableofcontents` command right after `\begin{document}`.

   Our code shall now look like this:

```
\documentclass[a4paper,12pt]{book}
\usepackage[english]{babel}
\usepackage{blindtext}
\usepackage[a4paper, inner=1.5cm, outer=3cm, top=2cm,
bottom=3cm, bindingoffset=1cm]{geometry}
\begin{document}
\tableofcontents
\chapter{Exploring the page layout}
In this chapter we will study the layout of pages.
\section{Some filler text}
\blindtext
\section{A lot more filler text}
More dummy text will follow.
\subsection{Plenty of filler text}
\blindtext[10]
\end{document}
```

4. Compile the code twice. Afterward, the first page of your output will contain this table:

Contents

Figure 3.15 – Table of contents

The `\tableofcontents` command tells LaTeX to produce and print a table of contents. During a typesetting run, LaTeX writes the headings into an auxiliary file with the `.toc` filename extension. The `\tableofcontents` command reads that `.toc` file in for printing the table of contents.

The LaTeX typesetting process is linear; it runs from the start to the end of the code. The `\tableofcontents` command comes at the beginning, and the headings come later. That's why we had to typeset twice:

1. In the first run, `\tableofcontents` did not know any headings, and the table of contents stayed empty. While continuing to run, LaTeX put the headings into the `.toc` file.
2. In the second run, `\tableofcontents` found and read the `.toc` file to print the contents.

It's good to keep that in mind for later: when you change a heading and compile the document, you can see the change in the text. But the table of contents will get this change in the next compiler run.

The sectioning commands create the table of contents entries. We used `\chapter`, `\section`, and `\subsection`, and we've got an entry for each.

A heading might be very long; it could span two or more lines. In that case, we might wish to shorten its corresponding table of contents entry. Let's see how.

We can use the optional arguments of the `section` commands to produce shorter entries, different from the actual headings. Let's edit the example shown in *Figure 3.15* by inserting shorter titles in square brackets:

```
\chapter[Page layout]{Exploring the page layout}
\section[Filler text]{Some filler text}
\section[More]{A lot more filler text}
\subsection[Plenty]{Plenty of filler text}
```

Compile the example twice. You will see that the headings stay the same, but the table of contents has changed:

Contents

Figure 3.16 – Shortened table of contents entries

Besides the mandatory argument producing the heading, each sectioning command understands an optional argument. If an optional argument is given, it will be used instead of the mandatory heading for the contents table entry.

In *Chapter 8, Managing Contents, Indexes, and Bibliography* we shall take a further look at this and learn how to customize the table of contents further. Let's review the sectioning commands for the `book`, `report`, and `article` classes again. There are seven levels in those base classes:

- `\part`: This is for dividing the document into major units. The numbering of other sectional units is independent of `\part`. A part heading will use a whole page in `book` and `report` documents.
- `\chapter`: This gives a large heading that will start at a new page, available in the `book` class and in the `report` class.
- `\section`, `\subsection`, and `\subsubsection`: These give bold headings, and they are available in all three classes.
- `\paragraph` and `\subparagraph`: Also available in all three classes, they produce a run-in heading. That means the heading runs straight into the text; there's no line break between the heading and the following text. Also, it's a sectioning command, and should not be confused with common text paragraphs.

Except `\part`, all sectioning commands reset the counter of the section that's one level below in the hierarchy. For instance, `\chapter` resets the section counter. This way, the sections will be numbered per chapter.

To sum up, such sectioning commands are easy to use, and they do a lot:

- `\part` and `\chapter` cause a page break before the heading
- All generate a number and a presentation for it, some depending on the higher-level counters (for example, *Section 1* of *Chapter 2* would generate `2.1`)
- Except `\part`, they reset the counter of the next-level sectional unit so that the lower-level unit will start with `1`
- They produce a table of contents entry
- They format the heading, usually in bold, and the larger it is, the higher it is in the hierarchy
- They save headings internally for using them in a page header

All sectioning commands provide a starred form, as follows:

```
\section*{title}
```

If you use this form, the numbering will be suppressed, and there won't be an entry in the table of contents or in a header. Look at the **Contents** heading in our example; this has actually been typeset by `\chapter*` inside the `\tableofcontents` macro.

Summary

In this chapter, we worked out how to design the overall layout of a document.

Specifically, we covered choosing page dimensions, margins, and orientation. We know how to switch to a two-column layout and how to adjust line spacing. Furthermore, we can now customize headers and footers, add footnotes, and include a table of contents in our document.

Beyond layout, we covered some general topics, such as changing document properties by selecting document class options and package options, as well as redefining existing commands.

Now that we've established the foundation, it's time to structure the content itself. In the next chapter, we ´ll learn how to create lists to present information in a clear and organized way.

4

Creating Lists

In the previous chapters, we focused on formatting, both at the text level and across entire pages. Now, we turn to structuring content more precisely. Over the following two chapters, we'll discuss how to organize information using lists and tables. We'll begin with lists.

Lists are an effective, reader-friendly way to present information. You can highlight key points in a format that's easy to scan and understand. The three most common list types are the following:

- **Bulleted lists**: Used to emphasize distinct points that stand out from the main text
- **Numbered lists**: Ideal for presenting points in a specific order
- **Definition lists**: Used to explain terms or concepts in a structured manner

In this chapter, you'll learn how to create each of these list types. We'll cover the following:

- Building lists
- Customizing lists

We'll start by creating basic lists and then move on to customizing their appearance and structure.

Building lists

We'll begin with unordered lists using bullet points to organize items. Next, we'll deal with ordered lists that use numbers or letters to indicate a sequence. Finally, we'll look at definition lists to present keywords alongside their corresponding explanations.

Creating a bulleted list

Let's start with the most basic type of list: a bulleted list. It contains just the items without numbers, where each item is marked with a bullet symbol. That's a much clearer way to present key points than squeezing them into a lengthy sentence within text in a paragraph.

As an example, we'll create a list of layout-related LaTeX packages introduced in the previous chapter. Follow these steps:

1. Create a new document with some introductory text:

```
\documentclass{article}
\begin{document}
\section*{Useful packages}
LaTeX provides several packages for designing the layout:
```

2. Now add the list, using an `itemize` environment and `\item` commands:

```
\begin{itemize}
  \item geometry
  \item typearea
  \item fancyhdr
  \item scrpage-scrlayer
  \item setspace
\end{itemize}
```

3. That was easy. Now we can end the document:

```
\end{document}
```

4. Click **Typeset** and have a look at the output:

Useful packages

LaTeX provides several packages for designing the layout:

- geometry
- typearea
- fancyhdr
- scrpage-scrlayer
- setspace

Figure 4.1 – A bulleted list

We started with a section heading followed by some introductory text. To create the list itself, we used an environment called `itemize`. As you've already learned about environments in *Chapter 2, Formatting Text and Creating Macros*, `\begin{itemize}` starts it and `\end{itemize}` ends it. The `\item` command tells LaTeX to add a new item to the list. The `\item` works only within a list environment. Each item can include text of any length and even span multiple paragraphs. Well, that's pretty easy.

When a list grows longer, we could improve clarity by dividing it. We can create lists under a list. It's advisable to use different bullet points to distinguish between the list levels easily. LaTeX does this for us automatically.

To illustrate this with our example, let's enhance our package list by grouping related items into topic categories. Here's how:

1. Modify the `itemize` environment of our example in the following way: create an `itemize` list for each topic and include it within an `\item`. The updated code looks like this:

```
\begin{itemize}
  \item Page layout
    \begin{itemize}
      \item geometry
      \item typearea
    \end{itemize}
  \item Headers and footers
    \begin{itemize}
      \item fancyhdr
      \item scrpage-scrlayer
    \end{itemize}
  \item Line spacing
    \begin{itemize}
      \item setspace
    \end{itemize}
\end{itemize}
```

   Note that we carefully closed each environment.

2. Compile the document to see the new list:

- Page layout
  - geometry
  - typearea
- Headers and footers
  - fancyhdr
  - scrpage-scrlayer
- Line spacing
  - setspace

Figure 4.2 – A bulleted list with two levels

Here, we placed lists within an `\item` point inside a list, creating a **nested list**. LaTeX supports up to four levels of nesting. If we exceeded this limit, LaTeX would raise an error:

```
! LaTeX Error: Too deeply nested.
```

The following are defaults:

- First-level items use bullet points
- Second-level items use wide dashes
- Third-level items are marked with an asterisk symbol (*)
- Fourth-level items use centered dots

Deeply nested lists are rarely used; such complicated structures might be difficult to follow. If your list becomes too complex, consider restructuring or dividing it.

In our example's source code, we indented each line inside the `itemize` environment. When there's another `itemize` environment within a surrounding `itemize` environment, the `\item` lines are even more indented. This makes it easy to identify the nesting level at a glance. While LaTeX doesn't require indentation, proper indentation within environments helps maintain the code, as it significantly improves its readability and structure. You can quickly see where environments begin and end. Indenting source code lines within an environment is a very good habit in general. You can also indent code lines to indicate that they belong to some parent line, like with an `\item` point that spans multiple lines, which we will see in the next section.

Using spaces or tabs to indent source code enhances the readability of the code. That has no impact on the output, since LaTeX treats multiple whitespace characters in a code line like a single space.

In the next section, we'll see how to list key points in a specific sequence and number them.

Building a numbered list

Bulleted lists work well when the order of the items doesn't matter. However, if the order is essential, we could organize the items by giving them numbers and creating a sorted list. This way, you guide the reader to follow the logical progression easily.

Let's prepare a short step-by-step tutorial for setting up the page layout using a numbered list. Follow these steps:

1. Open a new document and enter the following code:

```
\documentclass{article}
\begin{document}
\begin{enumerate}
  \item State the paper size by an option to the
        document class
  \item Determine the margin dimensions using one
        of these packages:
    \begin{itemize}
      \item geometry
      \item typearea
    \end{itemize}
  \item Customize header and footer by one
        of these packages:
    \begin{itemize}
      \item fancyhdr
      \item scrpage-scrlayer
    \end{itemize}
  \item Adjust the line spacing for the whole document
    \begin{itemize}
      \item by using the setspace package
      \item or by the command
            \verb|\linespread{factor}|
```

```
  \end{itemize}
\end{enumerate}
\end{document}
```

2. Click on **Typeset** to compile to generate these instructions:

    1. State the paper size by an option to the document class
    2. Determine the margin dimensions using one of these packages:
        - geometry
        - typearea
    3. Customize header and footer by one of these packages:
        - fancyhdr
        - scrpage-scrlayer
    4. Adjust the line spacing for the whole document
        - by using the setspace package
        - or by the command `\linespread{factor}`

Figure 4.3 – A numbered list with bulleted lists

We used an `enumerate` environment in a very similar way to the `itemize` environment, with each list entry beginning with the `\item` command. The main difference is that each `\item` line in our `enumerate` environment is automatically numbered instead of just having a bullet point in front. Again, we nested two lists, this time demonstrating that we can do so even when the lists are of different types. Mixed nesting can extend beyond four levels, but four is the maximum for each type of list, and a total of six levels is allowed for mixed lists.

The default numbering style for the `enumerate` environment is as follows:

- **First level**: 1., 2., 3., 4., ...
- **Second level**: (a), (b), (c), (d), ...
- **Third level**: i., ii., iii., iv., ...
- **Fourth level**: A., B., C., D., ...

The `\item` command can also take an optional argument. If you write `\item[text]`, LaTeX will print `text` instead of the default number or a bullet. This way, you could use any numbering and any symbol to replace the bullet.

Now that we've covered bulleted and numbered lists, let's look at a list type that we can use to typeset descriptions of multiple items.

Producing a definition list

We'll now proceed to the third type of list, namely, **definition lists**, also known as **description lists**. In this format, each list item begins with a term or phrase, followed by its corresponding explanation.

To illustrate this, we'll revisit our familiar theme of LaTeX packages. This time, instead of just listing the packages, we'll provide a brief description of each one. Let's pick a few from `https://ctan.org/topic/list`, which is a collection of list-related packages. This will also serve us in the next section, *Customizing lists*, where we'll work with the same packages to demonstrate customization features.

Let's begin by writing a brief summary of what each package can do. Follow these steps:

1. We will use a `description` environment. Create a document with the following code:

```
\documentclass{article}
\begin{document}
\begin{description}
  \item[paralist] provides compact lists and list
    versions that can be used within paragraphs,
    helps to customize labels and layout.
  \item[enumitem] gives control over labels
    and lengths in all kind of lists.
  \item[mdwlist] is useful to customize description
    lists, it even allows multi-line labels.
    It features compact lists and the capability
    to suspend and resume.
  \item[desclist] offers more flexibility in
    definition list.
  \item[multenum] produces vertical enumeration in
    multiple columns.
\end{description}
\end{document}
```

2. Compile the document to get the definition list:

    paralist provides compact lists and list versions that can be used within paragraphs, helps to customize labels and layout.

    enumitem gives control over labels and lenghts in all kind of lists.

    mdwlist is useful to customize description lists, it even allows multi-line labels. It features compact lists and the capability to suspend and resume.

    desclist offers more flexibility in definition list.

    multenum produces vertical enumeration in multiple columns.

    Figure 4.4 – A definition list

We used the `description` environment in much the same way as the other list environments, except that we used the optional argument of `\item` in square brackets. In the `description` environment, LaTeX automatically renders this argument in bold.

Compared to a bulleted list, the bullet symbols are replaced with bold keywords or terms.

We can also adjust the spacing of our lists, select bullet symbols, and modify the numbering style. Let's look at this in the next section.

Customizing lists

By default, LaTeX lists come with sensible settings for spacing, indentation, and symbols. However, in some cases, you may want to adjust the enumeration style, change the bullet symbols, or fine-tune the line spacing and indentation. Some packages help with space-saving and customizing the list's appearance. Let's begin by looking at the spacing.

Getting compact lists

A common question when writing with LaTeX is how to make lists take up less space. By default, LaTeX leaves generous spacing in and around lists, which can feel excessive. We'll now see how to reduce that.

Let's reduce the spacing in our list from *Figure 4.3*. We'll eliminate the extra space between the list items and around the entire list. Follow these steps:

1. In our numbered list example that produced *Figure 4.3*, include the `paralist` package and replace `enumerate` with `compactenum`, and `itemize` with `compactitem`:

```
\documentclass{article}
\usepackage{paralist}
```

```
\begin{document}
\begin{compactenum}
  \item State the paper size by an option to
        the document class
  \item Determine the margin dimensions using one
        of these packages:
  \begin{compactitem}
    \item geometry
    \item typearea
  \end{compactitem}
  \item Customize header and footer by one
        of these packages:
  \begin{compactitem}
    \item fancyhdr
    \item scrpage-scrlayer
  \end{compactitem}
  \item Adjust the line spacing for the whole document
  \begin{compactitem}
    \item by using the setspace package
    \item or by the command \verb|\linespread{factor}|
  \end{compactitem}
\end{compactenum}
\end{document}
```

2. Compile and compare the spacing:

    1. State the paper size by an option to the document class
    2. Determine the margin dimensions using one of these packages:
        - geometry
        - typearea
    3. Customize header and footer by one of these packages:
        - fancyhdr
        - scrpage-scrlayer
    4. Adjust the line spacing for the whole document
        - by using the setspace package
        - or by the command `\linespread{factor}`

Figure 4.5 – A compact list

3. Now extend the highlighted list item for `setspace` as follows:

```
\item by using the setspace package and one
      of its options:
  \begin{inparaenum}
    \item singlespacing
    \item onehalfspacing
    \item double spacing
  \end{inparaenum}
```

4. Compile and look at the change in the line spacing subject:

1. State the paper size by an option to the document class
2. Determine the margin dimensions using one of these packages:
   - geometry
   - typearea
3. Customize header and footer by one of these packages:
   - fancyhdr
   - scrpage-scrlayer
4. Adjust the line spacing for the whole document
   - by using the setspace package and one of its options: (a) singlespacing (b) onehalfspacing (c) double spacing
   - or by the command `\linespread{factor}`

Figure 4.6 – A list within a paragraph

The `paralist` package provides several alternative list environments designed to be typeset within paragraphs or in a very compact look. We loaded this package and replaced the standard environments with their compact counterparts, `enumerate` with `compactenum`, and `itemize` with `compactitem`. While the syntax is basically the same, the new environments don't produce additional vertical whitespace before and after a list. They also don't add vertical space around list items, resulting in a tighter layout. Lists and items are used with the same line spacing as regular text. Finally, it looks much more compact and saves space. In *step 3*, we used the new `inparaenum` environment, where the items are numbered but stay inline within the same paragraph.

For each standard list type, the `paralist` package offers three corresponding compact alternative environments.

For **bulleted lists**, it introduces the following environments:

- `compactitem`: A space-saving alternative to the `itemize` environment that removes vertical space before and after the list, and between its items
- `inparaitem`: A bulleted list formatted inline within a paragraph, rarely seen in print
- `asparaitem`: Each list item is formatted like a separate common LaTeX paragraph but starts with a bullet

For **numbered lists**, it provides the following list types:

- `compactenum`: A compact version of the `enumerate` environment, without extra vertical space before or after the list or between its items
- `inparaenum`: An enumerated list that appears inline within a paragraph
- `asparaenum`: Each list item is laid out like a regular LaTeX paragraph, but is numbered

For **description lists**, it adds the following types:

- `compactdesc`: A compact version of the `description` environment, eliminating the additional vertical space before and after the list and between its items
- `inparadesc`: A description list fits into a paragraph
- `asparadesc`: Each list item is formatted like a regular LaTeX paragraph, starting with the bold keyword or term, as in a `description` list

Now that we've adjusted the spacing, let's look at how to modify the bullet symbols and numbering styles.

Choosing bullets and numbering format

To meet language conventions or specific formatting guidelines, you might want to number list items using Roman numerals or letters, enclose them in parentheses, or follow them with dots. Some users also prefer using dashes instead of the default bullet points. The `enumitem` package offers flexible features to achieve all of this.

Let's work on this and change the numbering style. We will apply alphabetical numbering with circled letters. Furthermore, we'll change the default bullet symbol to a dash. To accomplish this, follow these steps:

1. We'll now use the `enumitem` package instead of `paralist`. We will use proper indentation within environments and return to the standard list syntax. However, to keep the layout compact, we'll add the `nosep` option to lists:

```
\documentclass{article}
\usepackage{enumitem}
\setlist{nosep}
\setitemize[1]{label=---}
\setenumerate[1]{label=\textcircled{\scriptsize\Alph*},
    font=\sffamily}
\begin{document}
\begin{enumerate}
  \item State the paper size by an option to the
        document class
  \item Determine the margin dimensions using one of
        these packages:
    \begin{itemize}
      \item geometry
      \item typearea
    \end{itemize}
  \item Customize header and footer by one of these
        packages:
  \begin{itemize}
    \item fancyhdr
    \item scrpage-scrheader
  \end{itemize}
  \item Adjust the line spacing for the whole document
  \begin{itemize}
    \item by using the setspace package
    \item or by the command \verb|\linespread{factor}|
  \end{itemize}
\end{enumerate}
\end{document}
```

2. Compile and check the output:

Ⓐ State the paper size by an option to the document class
Ⓑ Determine the margin dimensions using one of these packages:
 — geometry
 — typearea
Ⓒ Customize header and footer by one of these packages:
 — fancyhdr
 — scrpage-scrheader
Ⓓ Adjust the line spacing for the whole document
 — by using the setspace package
 — or by the command `\linespread{factor}`

Figure 4.7 – A customized enumerated list

We used commands from the `enumitem` package to customize list behavior. Let's take a closer look:

- `\setlist{nosep}`: The `\setlist` command applies properties valid for all types of lists. Here, we specified the `nosep` option to achieve very compact lists similar to the compact `paralist` environment. That setting removes all extra vertical spacing.
- `\setitemize[1]{label=---}`: The `\setitemize` command customizes bulleted lists. Here, we chose an em dash (`---`) as the label to get a wide leading dash.
- `\setenumerate[1]{label=\textcircled{\scriptsize \Alph*},font=\sffamily}`: The `\setenumerate` command customizes numbered lists. We used it to set a label and a font for the label. The `\Alph*` command means enumeration in capital letters.

We can also apply these options locally in square brackets. Other examples are as follows:

- `\begin{itemize}[noitemsep]` generates a compact bulleted list without any additional space between items and paragraphs
- `\begin{enumerate}[label=\Roman*.,start=3]` makes the list numbered by III., IV., and so on
- `\begin{enumerate}[label=\alph*)],nolistsep]` produces a very compact list numbered a), b), c), and so on

These are useful labeling commands for customizing numbered lists:

- `\arabic*`: 1, 2, 3, 4, ...
- `\alph*`: a, b, c, d, ...
- `\Alph*`: A, B, C, D, ...
- `\roman*`: i, ii, iii, iv, ...
- `\Roman*`: I, II, III, IV, ...

The asterisk * represents the current value of the list counter. You can freely add parentheses, punctuation, or other decorations. Later in this book, you'll see how to select from thousands of symbols for labels and bullets.

For convenience, there's also a short syntax: if you load the `enumitem` package with the `shortlabels` option, you may use a compact syntax such as `\begin{enumerate}[(i)]`, `\begin{enumerate}[(1)]` where `1`, `a`, `A`, `i`, and `I` correspond to `\arabic*`, `\alph*`, `\Alph*`, `\roman*`, and `\Roman*`, respectively. This allows for quick customization. However, consider using global commands to maintain consistent formatting throughout the document.

In this example, the `\verb|code|` command typesets code **verbatim**, "as it is" without interpreting it as LaTeX code. Instead of `|`, we can choose any character as the delimiter. Note that `\verb` cannot be used in arguments of commands such as `\section` and `\footnote`, nor in table headings. For longer blocks of such literal text, use the `verbatim` environment.

Let's move on to another helpful feature of the package.

Suspending and continuing lists

The `enumitem` package we used earlier also lets you continue list numbering. Let's build on the previous example to see how that works:

1. Just above the highlighted code line (`\item Adjust the line spacing...`), insert these lines to end the `enumerate` environment and continue with some regular text:

```
\end{enumerate}
\noindent\textbf{Tweaking the line spacing:}
```

2. Next, restart the `enumerate` environment with this line:

```
\begin{enumerate}[resume]
```

3. Leave the rest of the code, including the other items, unchanged. Compile the document to see the result:

Ⓐ State the paper size by an option to the document class
Ⓑ Determine the margin dimensions using one of these packages:
— geometry
— typearea
Ⓒ Customize header and footer by one of these packages:
— fancyhdr
— scrpage-scrheader
Tweaking the line spacing:
Ⓓ Adjust the line spacing for the whole document
— by using the setspace package
— or by the command `\linespread{factor}`

Figure 4.8 – A list that's resumed

In *step 1*, we temporarily ended the list and continued with regular text. In *step 2*, we restarted the list by using the `resume` option, which tells `enumitem` to continue the previous list with the next number. The starred version, `resume*`, would also use the options from the previous list, if given.

Examining the layout of lists

LaTeX provides sensible default layouts for lists; however, there may be occasions when you may want to modify this layout, such as adjusting the margins or the indentation of items. All layout dimensions are determined by LaTeX macros, so-called **lengths**.

There's a package that's great for visualizing layouts, which presents these length macros. It's called `layouts`. Let's use it to examine the dimensions of LaTeX's lists. We will use this short document:

```
\documentclass[12pt]{article}
\usepackage{layouts}
\begin{document}
\listdiagram
\end{document}
```

By simply compiling it, we will get the following diagram:

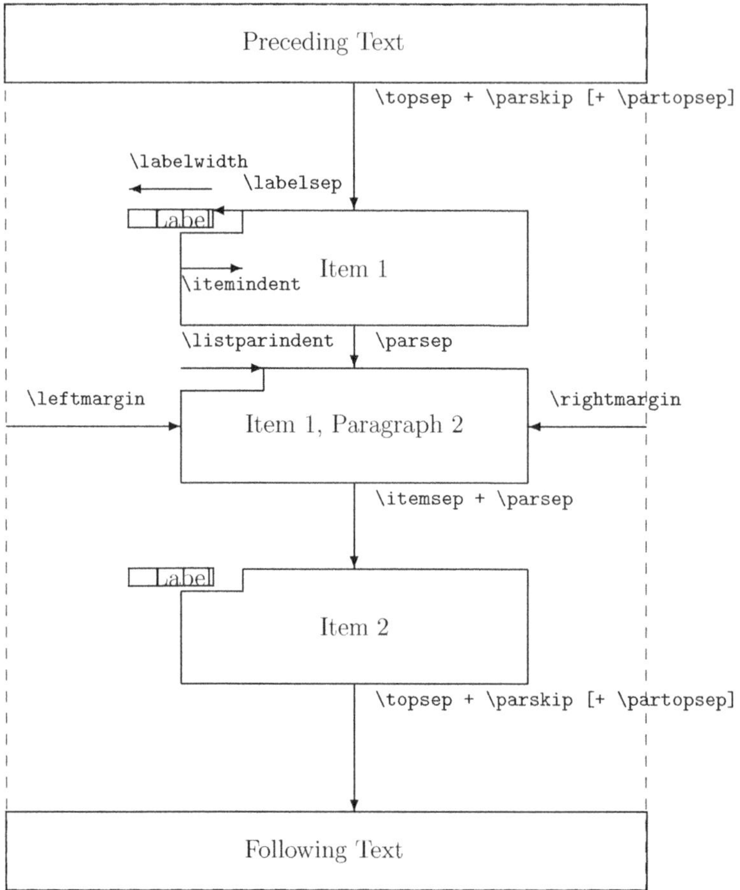

Figure 4.9 – The layout of lists

Isn't it impressive? The `layouts` package can do even more, which you can read about in its documentation at `https://texdoc.org/pkg/layouts` or by running `texdoc layouts` at the command line.

Use the LaTeX `\setlength` command to customize those lengths – for example, `\setlength{\labelwidth}{2cm}`. Applying them to individual lists and certain nesting depths is hard. If you need to modify the list layout, the `enumitem` package is helpful again. We can use its commands, such as `\setlist` and its `key=value` interface, to adjust the lengths, as shown in the previous diagram.

For example, if we would like to remove the space between list items in the description environment and to reduce the left margin, we could load the `enumitem` package and write the following:

```
\setdescription{itemsep=0cm,parsep=0cm,leftmargin=0.5cm}
```

Note, we don't use the backslash for keys. Similarly, `\setitemize`, `\setenumerate`, and `\setlist` can be used for fine-tuning. Try assigning values by yourself and test the effect on our examples. If you would like to learn more, have a look at the `enumitem` documentation at `https://texdoc.org/pkg/enumitem` or by running `texdoc enumitem` at the command prompt.

Summary

In this chapter, we explored lists as a means of organizing our content. Specifically, we covered how to create bulleted lists, numbered lists, and definition lists. Furthermore, we worked with compact and customized versions of such lists, including adjustments for spacing, as well as interrupting and resuming.

Such lists can help clarify and present your ideas more effectively. Next, we'll turn our attention to tables to typeset structured data in rows and columns.

5

Creating Tables

Scientific and technical documents often contain more than plain text; they also include structured data. In the previous chapter, we looked at ways to organize information with different kinds of lists and how to customize their layout. Now, we will turn to presenting data in tables.

We'll cover the following topics:

- Aligning text with tab stops
- Creating basic tables
- Improving table appearance
- Adding captions to tables
- Enhancing tables with additional packages

We'll start by simply arranging text in columns.

Aligning text with tab stops

Do you remember using tab stops on typewriters or in early word processing software for creating aligned columns of text? LaTeX offers a similar mechanism with the `tabbing` environment to neatly align text in columns.

Let's create a compact overview of LaTeX. We'll display one item per line, aligning key parts of the text, using the following steps:

1. Begin a new document and open a `tabbing` environment:

   ```
   \documentclass{article}
   \begin{document}
   \begin{tabbing}
   ```

2. Write the first line of text. Use \= to set tab stops and \\ to end the line:

```
\emph{Info:} \= Software \= : \= \LaTeX \\
```

3. Add more lines; use \> to jump to the next tab stop and again end the line with \\:

```
\> Author \> : \> Leslie Lamport \\
\> Website \> : \> www.latex-project.org
```

4. Close the `tabbing` environment and end the document:

```
\end{tabbing}
\end{document}
```

5. Click the **Typeset** button to compile the document, and take a look at the output:

Info: Software : LaTeX
Author : Leslie Lamport
Website : www.latex-project.org

Figure 5.1 – Simply aligned text

The `tabbing` environment automatically starts a new line. We used three basic commands to control alignment:

- \= sets a tab stop. You can place multiple tab stops on a single line. We typically do that in the first row.
- \\ ends a row.
- \> jumps to the next defined tab stop.

While tab stops are usually set in the first row, you can also redefine them in later lines using \=. For instance, if we already used two \> commands in a text line to navigate through two tab stops, inserting another \= command sets or adjusts a third tab stop on that line.

At `https://latexguide.org/tabbing`, you can see several examples using these tags.

The `tabbing` environment provides a quick way to create left-aligned columns. If the rows of the `tabbing` environment reached the end of a page, it would automatically continue on the next page. That makes it a basic yet efficient way to create tables that cross page breaks or even span multiple pages.

But what happens if the text exceeds the width and runs over the tab stop? Let's see how to handle that next.

In *Chapter 2, Formatting Text and Creating Macros*, we explored various font commands and declarations. You may recall a table that listed these commands alongside their output. Let's now create such a table ourselves, as follows:

1. Begin a new document, similar to *step 1* of our previous example, but this time, define a command for setting the header text in bold:

   ```
   \documentclass{article}
   \newcommand{\head}[1]{\textbf{#1}}
   \begin{document}
   \begin{tabbing}
   ```

2. Define tab stops in the first row using `\=` and use `\>` to jump to tab stops. Use the `\verb|...|` command you learned about in the previous chapter to display LaTeX code **verbatim**:

   ```
     \= \head{Command} \= \head{Declaration} \= \head{Example}\\
     \> \verb|\textrm{...}| \> \verb|\rmfamily| \> \rmfamily text\\
     \> \verb|\textsf{...}| \> \verb|\sffamily| \> \sffamily text\\
     \> \verb|\texttt{...}| \> \verb|\ttfamily| \> \ttfamily text
   ```

3. End the `tabbing` environment and the document:

   ```
   \end{tabbing}
   \end{document}
   ```

4. Compile the document and examine the output:

   Command Declaration Example
   \textrm{...\rmfamily text
   \textsf{...\sffamily text
   \texttt{...\ttfamily text

   Figure 5.2 – Overlapping aligned text

5. As we can see, the tab stops are too narrow at this point, causing the columns to overlap. We shall correct it. Create a new head row containing the tab stops, but this time, we will end the line with `\kill` to hide that line. Use filler text to specify the width between the tab stops, such as the longest text in the column. Complete it with further font commands. The tabbing code now looks like this:

   ```
   \begin{tabbing}
     \= \verb|\textrm{...}| \= \head{Declaration} \=
       \head{Example}\kill
   ```

```
  \> \head{Command} \> \head{Declaration} \> \head{Example}\\
  \> \verb|\textrm{...}| \> \verb|\rmfamily| \> \rmfamily text\\
  \> \verb|\textsf{...}| \> \verb|\sffamily| \> \sffamily text\\
  \> \verb|\texttt{...}| \> \verb|\ttfamily| \> \ttfamily text
\end{tabbing}
```

6. Compile again to get the final result:

Command	**Declaration**	**Example**
\textrm{...}	\rmfamily	text
\textsf{...}	\sffamily	text
\texttt{...}	\ttfamily	text

Figure 5.3 – Corrected aligned text

After realizing that our tab stops were too narrow, we added a new first row containing the tab stops. It consists of words representing the widest entries of each column. To keep this additional row from appearing in the output, we used the `\kill` command right at the end of the line; `\kill` at the end of a line skips printing the line but still registers the tab stops.

There are further useful commands that are handy if you frequently work with tabbed texts:

- `\+` at the end of a line (before `\\`) moves the left margin of the subsequent lines one tab stop to the right. Use it twice, `\+\+`, to move it two tab stops to the right, and so on.
- `\-` at the end of a line moves the left margin of the subsequent lines one tab stop to the left. Also, here, `\-\-` would move two tab stops to the left. `\-` is basically for canceling the effect of a previous indentation made by a prior `\+` command.
- `\<` cancels the effect of a preceding `\+` command for that line. It moves the left margin one tab stop to the left. We should only use it at the beginning of a line. Repeat it to move two tab stops to the left.

These commands give you solid control over the `tabbing` environment. For a full list of features, check the reference manual: `https://latex2e.org/tabbing`.

Inside a `tabbing` environment, declarations are local to the current item. A following `\=`, `\>`, `\\`, or `\kill` command would stop the effect.

Also, `tabbing` environments cannot be nested.

That wraps up how to align text in columns in a quick way. Next, let's see how to build tables with alignment and separation lines.

Creating basic tables

Sometimes we need more complex structures and formatting, such as centering within columns, inserting horizontal dividing lines, or even creating nested structures. LaTeX provides the `tabular` environment for building both simple and complex tables.

Let's recreate the table of font family commands from our previous example, this time with centered columns. We will also add some horizontal lines to visualize the table's border and the header. Follow these steps:

1. Create a new document. Define a command for setting the font for the head row:

```
\documentclass{article}
\newcommand{\head}[1]{\textnormal{\textbf{#1}}}
\begin{document}
```

2. Begin a `tabular` environment. As a mandatory argument, provide `ccc`, standing for three centered columns:

```
\begin{tabular}{ccc}
```

3. Write the table head row, use the `&` character to separate column entries, and finish with `\\` to end rows. Use `\hline` to insert horizontal lines:

```
  \hline
  \head{Command} & \head{Declaration} & \head{Output}\\
  \hline
```

4. Continue with the table body and end the environment and the document. To typeset LaTeX commands literally, write `\verb|\command|`:

```
  \verb|\textrm| & \verb|\rmfamily| & \rmfamily Example text\\
  \verb|\textsf| & \verb|\sffamily| & \sffamily Example text\\
  \verb|\texttt| & \verb|\ttfamily| & \ttfamily Example text\\
  \hline
\end{tabular}
\end{document}
```

5. Compile the document to see the table:

Command	Declaration	Output
\textrm	\rmfamily	Example text
\textsf	\sffamily	Example text
\texttt	\ttfamily	Example text

Figure 5.4 – A simple table

In *step 2*, we specified a list of characters in the mandatory argument. Each character defines how a column will be formatted. Since we used three characters, we've got three columns. The letter c stands for centered alignment, so the entries of all columns have been centered.

In *steps 3* and *4*, column entries are separated by an ampersand character (&), while \\ terminates a row. To improve readability, it's a good idea to align the ampersands, &, in the source code.

Within the column entries, you may use plain text as well as LaTeX commands. Just like in the `tabbing` environment, declarations are local to the current cell.

Furthermore, the `tabular` environment has an optional positioning argument just like the `minipage` environment. The full definition looks like this:

```
\begin{tabular}[position]{column specifiers}
  row 1 col 1 entry & row 1 col 2 entry ... & row 1 col n entry\\
  ...
\end{tabular}
```

The optional `[position]` argument sets vertical alignment of the entire table:

- `t` aligns at the top row of the table
- `b` aligns at the bottom row
- `c` centers the table vertically, which is the default

This optional argument may come in handy if you would like to place two tables next to each other or beside an image or other text.

In the following sections, we will discuss ways to customize tables, such as drawing lines; aligning to the left, right, or center; and merging cells across multiple columns or rows.

Drawing lines in tables

Within a tabular environment, we can use three types of lines:

- \hline draws a horizontal line spanning the entire width of the table.
- \cline{m-n} draws a horizontal line starting at column m and ending at column n. Both numbers are required: even if this line should cover only one particular column, for example, just column 3, we must write it as \cline{3-3}.
- \vline draws a vertical line across the entire height of the current row.

We'll make use of the \hline command in the upcoming sections.

Understanding formatting arguments

There are many more formatting options. Have a look at this example table, where we add l, c, r, and p as column specifiers, to demonstrate each kind of alignment:

```
\begin{tabular}{|l|c|r|p{1.7cm}|}
  \hline
  left & centered & right & a fully justified paragraph cell\\
  \hline
  l & c & r & p\\
  \hline
\end{tabular}
```

This code generates the following table:

left	centered	right	a fully justified paragraph cell
l	c	r	p

Figure 5.5 – A table with different alignments

The options understood by the tabular environment are as follows:

- l aligns content to the left.
- c centers content.
- r aligns content to the right.

- p{width} is for a "paragraph" cell of a fixed width. If you place several p columns next to each other, their cells will be aligned at their top line. It behaves like using \parbox[t]{width} inside a cell.
- @{code} inserts code instead of empty space before or after a column. The code can also be some text, or it could be left empty to suppress this space entirely, such as @{}. We'll make frequent use of this later in this section.
- *{n}{options} repeats the specified options n times. n is a positive integer, and the options may consist of one or more column specifiers, including * as well.
- | draws a vertical line along the column.

It's best to avoid using vertical lines in tables. Lines should provide subtle visual guidance without interfering with readability. Reading rows from left to right is easier without separation lines. Many scientific writing guidelines recommend this approach, and some journals even require it.

The array package offers many further features. After loading it with \usepackage{array}, you may use the following options to tabular:

- m{width} is similar to \parbox{width}: the baseline is in the middle.
- b{width} behaves like \parbox[b]{width}: the baseline is at the bottom.
- !{code} inserts code. In contrast to @{...}, the space between columns will not be suppressed.
- >{code} can be used before an l, c, r, p, m, or b option, and inserts code right at the beginning of each entry of that column. We'll use this feature frequently to insert font commands.
- <{code} can be used after an l, c, r, p, m, or b option, and inserts code directly at the end of the entry of that column.

This example shows loading the array package and demonstrates the effect of @{} and the p, m, and b alignment arguments:

```
\documentclass{article}
\usepackage{array}
\begin{document}
\begin{tabular}{@{}lp{1.2cm}m{1.2cm}b{1.2cm}@{}}
  \hline
  baseline & aligned at the top & aligned at the middle
  & aligned at the bottom\\
```

```
  \hline
\end{tabular}
\end{document}
```

The output table is as follows:

baseline	aligned at the top	aligned at the middle	aligned at the bottom

Figure 5.6 – A table with different vertical alignments

Notice the last column in *Figure 5.6*; the text isn't visually aligned with the bottom of the table cell. That's important to understand; the option `b` means that the baseline of the cell text shall be the bottom line. Then, LaTeX aligns all baselines at the same vertical level. So, to align text where the baseline is its bottom line with the other baselines, the text must shift upward. You can think of baselines as anchor lines; all cells align their content based on these shared anchor points.

Improving table appearance

Let's now look at tweaking details and merging cells.

Increasing the row height

You might have noticed that horizontal lines in a table can come uncomfortably close to capital letters. To address this, the `array` package introduces a length called `\extrarowheight`. It's `0` by default, but if you set it to a positive value, this adds extra vertical space above every row of the table.

The next example, following the very first example of this chapter, shows how to increase the row height in the highlighted line. It also shows the effect of further `array` formatting options, as follows:

```
\documentclass{article}
\usepackage{array}
\setlength{\extrarowheight}{4pt}
\begin{document}
\begin{tabular}{@{}>{\itshape}ll!{:}l<{.}@{}}
  \hline
  Info:     & Software & \LaTeX\\
```

```
              & Author   & Leslie Lamport\\
              & Website  & www.latex-project.org\\
  \hline
\end{tabular}
\end{document}
```

The output is as follows:

Info:	Software	:	LaTeX.
	Author	:	Leslie Lamport.
	Website	:	www.latex-project.org.

Figure 5.7 – A stretched table

Here, we applied `>{\itshape}` to change the font of a column to italics. The `>{...}` construct is often used to insert an alignment declaration such as `\centering`. However, there's a subtle issue: declarations such as the `\centering` command might change the internal meaning of `\\`, which is a shortcut for `\tabularnewline` within tables. But the `array` package offers a command to repair it. In such cases, just add the `\arraybackslash` command, as in the following example:

```
\begin{tabular}{>{\centering\arraybackslash}p{5cm}}
```

Without this, the text in paragraph cells stated by `p`, `m`, or `b` will stay fully justified.

You can also insert extra vertical space after a specific row using an optional argument for `\\`, such as `\\[10pt]`.

You may even stretch a whole table: the `\arraystretch` command contains a stretching factor with a default value of `1`. You can redefine it to increase it. For example, `\renewcommand{\arraystretch}{1.5}` will increase the height of the rows by a factor of *1.5*, which means adding 50 percent space. Unlike `\extrarowheight`, it adds space both above and below the row. To keep the effect limited, place the command inside a group or an environment. Alternatively, you can reset it to the default value with `\renewcommand{\arraystretch}{1}`.

Beautifying tables

Our tables may still not look as polished as those in professionally typeset books. In particular, we could improve the lines and their distances from the text content. The `booktabs` package helps with that. By using this package, you can greatly enhance the appearance of your tables with new commands that replace `\hline` and `\cline`.

Here's how:

1. Using our previous example from *Figure 5.4*, load the `booktabs` package:

```
\usepackage{booktabs}
```

2. Use the new `\toprule`, `\midrule`, and `\bottomrule` commands instead of `\hline`. You can specify the thickness as an optional argument. The table code becomes the following:

```
\begin{tabular}{ccc}
  \toprule[1.5pt]
  \head{Command} & \head{Declaration} & \head{Output}\\
  \midrule
  \verb|\textrm| & \verb|\rmfamily| & \rmfamily Example text\\
  \verb|\textsf| & \verb|\sffamily| & \sffamily Example text\\
  \verb|\texttt| & \verb|\ttfamily| & \ttfamily Example text\\
  \bottomrule[1.5pt]
\end{tabular}
```

3. Compile to see the difference:

Command	Declaration	Output
\textrm	\rmfamily	Example text
\textsf	\sffamily	Example text
\texttt	\ttfamily	Example text

Figure 5.8 – A stretched table

Especially in British typesetting, the term *rule* is commonly used to refer to a line. The `booktabs` developer adopted this terminology for the package's commands. These are their definitions:

- `\toprule[thickness]` draws a horizontal line at the top of the table. If desired, a `thickness` value may be specified, such as `1pt` or `0.5mm`.
- `\midrule[thickness]` creates a horizontal dividing line between table rows.
- `\bottomrule[thickness]` draws a horizontal line to finish off a table.
- `\cmidrule[thickness](trim){m-n}` draws a horizontal line from column `m` to column `n`. The `(trim)` argument is optional, similar to the `thickness` argument. It could be `(l)` or `(r)` to trim the line at its left or right end. Write `(lr)` to trim at both ends. You can even specify a trim width like in `(l{10pt})`.

The booktabs package intentionally avoids defining vertical lines. Neither vertical nor double lines are recommended for clean, professional tables. In fact, both are generally regarded as poor typographic style. Instead, rely on \toprule, \midrule, and \bottomrule. You can use them without optional arguments – let's look at how to do that next.

Customizing line thickness and spacing

We previously mentioned the \setlength command in the *Increasing the row height* section of this chapter. Rather than specifying a line thickness with an optional argument to \toprule, \midrule, \cmidrule, or \bottomrule, we can always omit it. Instead, we set it globally, once for our whole document, by using \setlength in the preamble.

So, for example, after \usepackage{booktabs}, you can write the following:

```
\setlength{\heavyrulewidth}{1.5pt}
```

Now just use \toprule and \bottomrule without an argument, and they will always be 1.5 pt thick.

The following booktabs package lengths can be adjusted:

- \heavyrulewidth: The thickness of top and bottom lines with \toprule and \bottomrule
- \lightrulewidth: The thickness of middle lines with \midrule
- \cmidrulewidth: The thickness of \cmidrule
- \cmidrulekern: Trimming in \cmidrule
- \abovetopsep: The space above the top rule; the default is 0pt
- \belowbottomsep: The space below the bottom rule; the default is 0pt
- \aboverulesep: The space above \midrule, \cmidrule, and \bottomrule
- \belowrulesep: The space below \midrule, \cmidrule, and \toprule

Feel free to experiment with these settings. The default values are well balanced, but you may change them. Adjustments made in your preamble will apply to all tables in your document.

Spanning entries across multiple columns

To group related columns under a shared heading, we can merge table cells using the `\multicolumn` command.

In our example, both commands and declarations represent input, while the third column contains output. To reflect this structure, we'll revise the header row to span columns as follows:

1. Based on our previous example, insert a new header row. Use `*{3}l` to define three left-aligned columns. Put `@{}` before and after to remove inter-column space. Apply the `\multicolumn` command to merge cells. Adjust the column formatting argument and the middle rule. The changes are highlighted here:

```
\begin{tabular}{@{}*{3}l@{}}
  \toprule[1.5pt]
  \multicolumn{2}{c}{\head{Input}} &
  \multicolumn{1}{c}{\head{Output}}\\
  \head{Command} & \head{Declaration} & \\
  \cmidrule(r){1-2}\cmidrule(l){3-3}
  \verb|\textrm| & \verb|\rmfamily| & \rmfamily Example text\\
  \verb|\textsf| & \verb|\sffamily| & \sffamily Example text\\
  \verb|\texttt| & \verb|\ttfamily| & \ttfamily Example text\\
  \bottomrule[1.5pt]
\end{tabular}
```

2. Compile and examine the output:

Input		Output
Command	Declaration	
\textrm	\rmfamily	Example text
\textsf	\sffamily	Example text
\texttt	\ttfamily	Example text

Figure 5.9 – A table with merged cells

We used the `\multicolumn` command twice: once to merge two cells and, surprisingly, another time just for one cell – we'll see in a second why. Let's first look at its definition:

```
\multicolumn{number of columns}{formatting options}{entry text}
```

The number of columns to be spanned may be a positive integer or just 1. The formatting options will be applied instead of the options specified in the `tabular` definition for this cell.

We took advantage of this when we used `\multicolumn{1}{c}{…}`, overriding the `l` option of the column with a `c` option to get just this cell centered.

The other change we made concerns `\cmidrule`. We used it instead of `\midrule`, together with the trimming option, to get a gap between the input and the output column for improving visual clarity.

Inserting code column-wise

There are many more font commands that we would like to add to the table. However, writing `\verb|…|` commands in each cell would quickly become tedious. Instead, we can take advantage of the `>{…}` construct from the `array` package to apply formatting to entire columns.

Let's update our `tabular` definition to set our input columns in the typewriter font. We'll also add a new column on the left to describe the command's type. Let's go:

1. Extend the preamble of our example by defining a `\normal` macro. It will use the `\multicolumn` command to generate an `l` cell, no matter what the column formatting is:

```
\documentclass{article}
\usepackage{array}
\usepackage{booktabs}
\newcommand{\head}[1]{\textnormal{\textbf{#1}}}
\newcommand{\normal}[1]{\multicolumn{1}{l}{#1}}
\begin{document}
```

2. As the `\verb` command can't be used in table headers, we will use `\ttfamily` to get the typewriter font. We'll also use `\textbackslash` here to avoid repeating that long command phrase within the cells. Use `*{2}>{…}` to insert it twice. Then, add `<{Example text}` to the last column to save typing work:

```
\begin{tabular}{@{}l*{2}{>{\ttfamily\textbackslash}l}l%
  <{Example text}@{}}
  \toprule[1.5pt]
  & \multicolumn{2}{c}{\head{Input}} &
  \multicolumn{1}{c}{\head{Output}}\\
```

3. We'll use our `\normal` macro to avoid the typewriter formatting in the header:

```
& \normal{\head{Command}} & \normal{\head{Declaration}}
& \normal{}\\
\cmidrule(lr){2-3}\cmidrule(l){4-4}
```

4. Now we continue listing the font command names:

```
  Family & textrm & rmfamily & \rmfamily\\
  & textsf & sffamily & \sffamily\\
  & texttt & ttfamily & \ttfamily\\
  \bottomrule[1.5pt]
\end{tabular}
\end{document}
```

5. Compile, and then look at the result:

	Input		Output
	Command	Declaration	
Family	\textrm	\rmfamily	Example text
	\textsf	\sffamily	Example text
	\texttt	\ttfamily	Example text

Figure 5.10 – A table with column formatting commands

By writing `>{\textbackslash\ttfamily}l`, we defined a left-aligned column, where every entry automatically begins with a backslash and appears in typewriter font. We wrote `*{2}{...}` to define two columns of this style. Since the example text was inserted according to our table definition with `<{...}>`, we only needed to place the declarations in the last column without repeating the actual text.

Spanning content across multiple rows

We've seen how to span text across multiple columns. But what about spanning across rows? While LaTeX itself doesn't offer a built-in command for this, the `multirow` package provides exactly what we need.

Let's center the word "Family" vertically, spanning across three rows. Here's how:

1. In our previous example, additionally load the `multirow` package:

```
\usepackage{multirow}
```

2. Replace the word `Family` with `\multirow{3}{*}{Family}`:

```
\multirow{3}{*}{Family} & textrm & rmfamily & \rmfamily\\
```

3. Compile to see the small change:

	Input		Output
	Command	Declaration	
	\textrm	\rmfamily	Example text
Family	\textsf	\sffamily	Example text
	\texttt	\ttfamily	Example text

Figure 5.11 – Vertically merged cells

We used the `\multirow` command to span three rows. Its definition is as follows:

```
\multirow{number of rows}{width}{entry text}
```

The entry will span that number of rows from the row on which `\multirow` is used. If the number is negative, it will span the rows above.

You can specify a width or just write * to use the natural width. If you set a width, LaTeX would wrap the text to fit.

The `multirow` command understands further optional arguments for fine-tuning. The documentation at `https://texdoc.org/pkg/multirow` describes this.

Now that we've mastered table layout, let's look at how to add caption text.

Adding captions to tables

When working on longer documents where our text references the tables, it's helpful to add captions and numbers to our tables. Numbering the tables allows easy referencing, whereas captions give readers context about the content of the table. LaTeX has built-in support for both.

Let's now finalize our font table. We'll extend the table by listing additional font commands. We'll use the first column to describe the category of the font commands, such as family, weight, and shape. Then, we'll add another column to demonstrate the effect of combining font commands.

To finish, we shall center the table and provide a number and a caption. To do that, we will put a `table` environment around our example table, use the `\centering` command inside it, and insert a `\caption` command at the end of the `table` environment. We will add more font commands and add another column on the right, containing more examples. Let's break it down into the following steps:

1. Start with the `article` class, and load the `array`, `booktabs`, and `multirow` packages:

```
\documentclass{article}
\usepackage{array}
\usepackage{booktabs}
\usepackage{multirow}
```

2. Define a macro for formatting the header cells and a macro for normal cells, which we want to be left-aligned:

```
\newcommand{\head}[1]{\textnormal{\textbf{#1}}}
\newcommand{\normal}[1]{\multicolumn{1}{l}{#1}}
```

3. Begin the document:

```
\begin{document}
```

4. Now, we will create the table, center the content, and write up all the rows:

```
\begin{table}
  \centering
  \begin{tabular}{@{}l*{2}{>{\textbackslash\ttfamily}l}%
    l<{Example text}l@{}}
    \toprule[1.5pt]
    & \multicolumn{2}{c}{\head{Input}}
    & \multicolumn{2}{c}{\head{Output}}\\
    & \normal{\head{Command}}
    & \normal{\head{Declaration}}
    & \normal{\head{Single use}} & \head{Combined}\\
    \cmidrule(lr){2-3}\cmidrule(l){4-5}
    \multirow{3}{*}{Family} & textrm & rmfamily
    & \rmfamily & \\
    & textsf & sffamily & \sffamily& \\
    & texttt & ttfamily & \ttfamily& \\
    \cmidrule(lr){2-3}\cmidrule(lr){4-4}
```

```
    \multirow{2}{1.1cm}{Weight} & textbf & bfseries
    & \bfseries
    & \multirow{2}{1.8cm}{\sffamily\bfseries Bold and
      sans-serif}\\
    & textmd & mdseries & \mdseries & \\
    \cmidrule(lr){2-3}\cmidrule(lr){4-4}
    \multirow{4}{*}{Shape} & textit & itshape
    & \itshape & \\
    & textsl & slshape & \slshape &
    \multirow{2}{1.8cm}{\sffamily\slshape Slanted and
      sans-serif}\\
    & textsc & scshape & \scshape & \\
    & textup & upshape & \upshape  & \\
    \cmidrule(lr){2-3}\cmidrule(lr){4-4}
    Default & textnormal & normalfont & \normalfont & \\
    \bottomrule[1.5pt]
  \end{tabular}
  \caption{\LaTeX\ font selection}
\end{table}
```

5. End the document:

```
\end{document}
```

6. Compile, and our table is now ready:

	Input		Output	
	Command	Declaration	Single use	Combined
Family	\textrm	\rmfamily	Example text	
	\textsf	\sffamily	Example text	
	\texttt	\ttfamily	`Example text`	
Weight	\textbf	\bfseries	**Example text**	**Bold and sans-serif**
	\textmd	\mdseries	Example text	
Shape	\textit	\itshape	*Example text*	
	\textsl	\slshape	*Example text*	*Slanted and sans-serif*
	\textsc	\scshape	EXAMPLE TEXT	
	\textup	\upshape	Example text	
Default	\textnormal	\normalfont	Example text	

Table 1: LaTeX font selection

Figure 5.12 – A table with a caption

We wrapped the `tabular` environment inside a `table` environment. It's used in this way together with the `\caption` command:

```
\begin{table}[placement options]
  table body
  \caption{table title}
\end{table}
```

The `table` environment is a **floating environment**. Unlike regular text, it can appear somewhere else other than what is defined by its position in the source code. The optional `placement` argument determines where the table might appear. However, LaTeX will position a table within the text to achieve good page breaks without too much empty space at the end of a page. We will spend some time discussing floating environments in the next chapter, specifically in the context of graphic placement, and the same principles apply here. As we'll see with figures in the next chapter, `\begin{table}[htbp!]` is the most flexible choice.

`\caption` understands an optional argument as well. If you write `\caption[short text]{long text}`, then the short text will appear in a list of tables and the long text in the document body. That's useful if you need very long descriptive captions.

Tables are automatically numbered. Let's talk about the positioning and formatting of captions in the following two sections.

Placing captions above tables

In professional typesetting, it's very common to place captions above tables rather than below. We can achieve this by writing `\caption` before the table body. However, LaTeX assumes the caption will always be below, resulting in a cramped appearance of the table. There'll be too little space between the caption and the following table. So, you might wish to add some space, for instance, by entering `\vspace{10pt}` directly after a top caption.

Remember the `booktabs` package? If you begin tables with `\toprule`, just specify the `\abovetopskip` length, as in the following example:

```
\setlength{\abovetopsep}{10pt}
```

By putting this line into your preamble, a 10 pt space would be added below the caption and above the top line of the table.

Customizing captions

By default, captions look like regular body text; there's no visual difference. Would you like to have a slight change in font size, different formatting of the label, some margins or indentation, or any other customization? The `caption` package offers a powerful solution.

By using a few options, you can enhance the visual appearance of all of your captions. Try the following:

```
\usepackage[font=small,labelfont=bf,margin=1cm]{caption}
```

This configuration makes your captions slightly smaller than the regular text, displays the label in bold, and ensures it will not be as wide as normal text, adding a margin of 1 cm to each side of the caption. The package provides many customization features, both for document-wide settings and fine-tuning individual captions. Its documentation is extensive and helpful. Either visit `https://texdoc.org/pkg/caption` or type `texdoc caption` at the command line to learn more.

There are various packages for table layout and appearance. In the next section, we will get to know such packages.

Enhancing tables with additional packages

While creating tables in LaTeX, we may face various formatting challenges. For example, we may need to adjust column widths, insert page breaks within tables, apply colors, rotate tables, and achieve a specific alignment. In the following sections, we'll examine additional packages designed to address these needs.

You can find example tables and links to the documentation at `https://latexguide.org/tables` for each of the following sections.

Auto-fitting columns to the table width

`l`, `c`, and `r` columns adjust to the width of their content. For `p` columns, you set a fixed width. This way, it's hard to predict the actual width of the table. Wouldn't it be a good idea to specify the table width and let LaTeX decide how wide the columns may be? That's exactly what the `tabularx` package does. Use it as follows:

```
\usepackage{tabularx}
...
\begin{tabularx}{width}{column specifiers}
  ...
\end{tabularx}
```

The new `tabularx` environment takes one extra argument: the desired width of the table. It introduces a new column type, `X`. It behaves like a `p` column, but `X` columns automatically stretch to fill the available space. One `X` column would take all of the available space. If you use several `X` columns, they will share the space equally. So, you could write, for instance, the following:

```
\begin{tabularx}{0.6\textwidth}{lcX}
```

This way, you would get a table occupying 60 percent of the text width – a left-aligned and a centered column as wide as their content, and a paragraph column as wide as possible until 60 percent is reached.

Despite its simplicity, the `tabularx` documentation provides further examples, explains the derived types, and offers advice, such as avoiding `\multicolumn` entries that cross any `X` column. Read the documentation at `https://texdoc.org/pkg/tabularx` or run `texdoc tabularx` at the command line.

There are two related approaches. Standard LaTeX provides a starred version of the `tabular` environment:

```
\begin{tabular*}{width}[position]{column specifiers}
```

The table adjusts to `width` by modifying the inter-column space. `tabularx` has been developed to satisfy this need in a more useful way.

The `tabulary` package provides another sophisticated `tabular` environment that takes the total width. It weights each column width according to the natural width of the widest cell in the column.

In general, the `tabularx` package is an excellent choice for adjusting the table width to the text width and is probably the most popular tabular package.

Generating multi-page tables

All the `tabular` environments that we've seen so far can't extend across page boundaries. The `tabbing` environment is an exception, since it works differently.

When a table contains a large amount of data, we need another approach. There are several packages that offer different ways to handle it:

- `longtable` provides an environment with the same name that's like a multi-page version of `tabular`. It provides commands to set table captions, continued captions, and special headers and footers when a page break occurs. It's probably the easiest way for multi-page tables, and therefore it's the most popular. The package documentation describes all you need. In combination with the `booktabs` package, you will get excellent results. This is the most commonly used package.
- `ltxtable` provides a combination of `longtable` and `tabularx`.
- `ltablex` is another approach to combine the features of `longtable` and `tabularx`.
- `supertabular` offers another multi-page extension of the internally used `tabular` environment, providing optional table tails and heads where page breaks occur. It's recommended for two-column documents.
- `xtab` extends `supertabular` and reduces some of its weaknesses.
- `stabular` implements a simple way to use page breaks in `tabular` without much fuss.

If you need tables to be able to extend across multiple pages, you can check out the packages' manuals at `https://texdoc.org/`, or use the `texdoc` command on your computer. Among them, the `longtable` package is probably the most popular choice.

Coloring tables

We discussed coloring text in *Chapter 2, Formatting Text and Creating Macros*. To color tables, use the `colortbl` package. We can combine all this by using the `table` option with the `xcolor` package:

```
\usepackage[table]{xcolor}
```

The package allows coloring columns, rows, single entries, and lines in many ways. The package documentation can tell you more; find it at `https://texdoc.org/pkg/xcolor`.

Using landscape orientation

If a table is too wide to fit in portrait mode, you can rotate it into landscape orientation. The `rotating` package offers an environment called `sidewaystable`, which behaves like the regular `table` environment but rotates the entire table and its caption by 90 degrees, placing it on its own page. The package includes further rotation-related environments and commands.

Aligning columns at the decimal point

Columns containing numbers are more readable when the entries are aligned at the decimal point or an exponent if present. Several packages assist with this:

- `siunitx` is a very modern and actively developed package primarily intended for typesetting values with units in a consistent way, according to scientific conventions. It provides a tabular column type for such decimal alignment of numbers.
- `dcolumn` offers a column type for aligning at a comma, a period, or another specific character.
- `rccol` defines a column type where numbers are right-centered: they are centered relative to other entries but right-aligned with each other. This way, corresponding digits are aligned along the column. `siunitx` would be the better choice, however.

In contrast to `dcolumn` and `rccol`, the `siunitx` package is newer and very powerful.

Dealing with narrow columns

Text in very narrow columns might need special attention because justification can be difficult if there's limited space. Here's some advice:

- Ensure correct hyphenation. If necessary, improve it as we did in *Chapter 2, Formatting Text and Creating Macros*. TeX doesn't hyphenate the first word of a line, a box, or a table entry. So, a long word may cross the column boundary. To enforce hyphenation, insert an empty word: write `\hspace{0pt}` directly at the beginning.
- Load the `microtype` package to improve justification. It has a significant effect, especially in narrow columns.
- Full justification in `p` columns and the like may look bad because big gaps may appear. Consider using `>{\raggedright\arraybackslash}` for such columns.
- From the `ragged2e` package, using the `\RaggedRight` command can give even better results and doesn't require `\arraybackslash`.

These strategies help prevent unwanted gaps and improve the overall appearance of your tables.

Summary

In this chapter, you've learned how to create and format tables. We covered methods for organizing text into columns, adding captions, merging rows and columns, using packages to fit columns automatically, and creating colored, landscape, and even multi-page tables.

You can open the documentation for any mentioned package by running texdoc packagename at the command line or by visiting https://texdoc.org/pkg/<packagename>.

LaTeX can also generate a list of tables, such as a table of contents. We'll look at that in *Chapter 8, Managing Contents, Indexes, and Bibliography*.

LaTeX numbers our tables automatically. We can use these numbers to reference the tables. *Chapter 7, Using Cross-References*, is dedicated to referencing sections and objects, including tables.

The same principle applies to figures with images, which is the topic of the next chapter.

6

Including Images

In the previous chapter, we organized data using tables. However, not all information fits into rows and columns; sometimes, images communicate it better, such as diagrams, illustrations, screenshots, and photos. In this chapter, we'll learn how to include images as figures, add captions, and reference them just like tables. We'll also take a close look at how to control the placement of floating figures, a concept that applies to tables as well.

We will cover the following topics:

- Choosing the right image format
- Creating images
- Embedding images
- Scaling an image
- Including entire pages
- Putting images behind the text
- Managing floating figures
- Arranging multiple images
- Letting text flow around images

By the end of this chapter, you'll be able to control the placement of images within your document with confidence and accuracy.

We will start by looking at the graphics formats.

Choosing the right image format

Let's quickly look at the supported image file formats: **EPS**, **PDF**, **JPG**, and **PNG**. **PS** stands for **PostScript**, while **EPS** stands for **Encapsulated PostScript**. You're already familiar with PDF, which stands for **Portable Document Format**. PDF files don't have to be full pages; they can also be small images. And you know the image formats PNG and JPG, which are widely used for screenshots and photos.

If your image is already in one of these four formats, you can include it directly in your document, as these formats are supported by default. Converting between formats won't improve quality. However, when you create or export an image, you can still choose the most suitable file format. Here's how to decide.

EPS and **PDF** are both **vector graphics formats**. They are scalable and maintain high quality even at high resolution or when zoomed in. Therefore, whenever possible, you should prefer PDF (or EPS), for example, when exporting drawings or diagrams from other office software. Vector formats are typically used to achieve the best quality while keeping file sizes small.

PNG and JPG are **bitmap formats**, also known as **raster graphics**, and are commonly used for photographs and screenshots. When you zoom into bitmap images, they may lose sharpness as individual pixels become more visible, leading to blocky or blurred edges. PNG images use lossless compression, while JPG images may degrade in quality during saving. So, if you make screenshots, it's best to use PNG; or, if you choose JPG, make sure no lossy compression is applied: usually, you can choose a quality level when saving. For photos, the JPG format is a good choice to keep the resulting PDF file size small because its compression works well with natural images that have soft color transitions. You'll rarely notice a loss in photo quality, while the same compression can easily distort sharp lines in diagrams or screenshots.

In addition to supporting vector graphics, both EPS and PDF can also contain bitmap images. They are also referred to as **container formats** in general because they can embed multiple types of content within a single file, such as images, text, fonts, and multimedia content. We focus on single images here.

There are numerous tools available to convert between graphic formats. The following three programs are popular, and both **TeX Live** and **MiKTeX** include them:

- `dvips` converts DVI (`.dvi`) files to PostScript (`.ps`) format
- `ps2pdf` converts PostScript files to PDF
- `epstopdf` converts EPS (`.eps`) files to PDF

These are command-line tools. Some LaTeX editors integrate them to provide single-click compiling, for example, from `.tex` to `.dvi`, then to `.ps`, and then to `.pdf`.

Since 2011, **pdflatex** calls `epstopdf` automatically when you include EPS images to convert them on the fly.

Creating images

Let's briefly look at some powerful, free, and open source tools for creating, processing, and converting graphics you can include in your documents. All are available for Windows, macOS, and Linux. On Linux, check your app store or software repository first to see whether the program is available for your specific system version before downloading it from the official website.

Producing and processing bitmap images

The following tools are especially recommended. All of them support many standard formats, including PNG, JPG, and GIF, and can export to PDF:

- **GIMP** (`https://www.gimp.org`) is an **image editor** often compared to Adobe Photoshop. It's ideal for photo editing and retouching tasks. You can use it to crop, resize, sharpen, adjust colors, or combine multiple images. GIMP supports many standard formats, including PNG, JPG, and even Photoshop's PSD files.
- **Pinta** (`https://www.pinta-project.com`) is an **image editor** inspired by Paint.NET, which is also free but available only for Windows. Compared to GIMP, Pinta has fewer features but uses far less memory and disk space. Its simpler design also makes it much easier to use.
- **Krita** (`https://krita.org`) is a full-featured **digital art studio**. You can paint with realistic brushes and layers and use it on tablets, making it great for digital illustrations.

- **MyPaint** (`https://www.mypaint.app`) is a lightweight, fast **painting program** that's ideal for quick sketches and tablet drawing.
- **ImageMagick** (`https://imagemagick.org`) is a **software suite** for creating, editing, and converting images. Unlike graphical editors, it's mainly used from the command line or in scripts to automate image processing and batch tasks. You can use it to crop, resize, rotate, adjust colors, or apply effects to images. It supports hundreds of file formats, making it especially useful for converting between them.

These tools cover most needs for working with bitmap images.

Editing vector graphics

The following tools use vector shapes, which keep your graphics sharp at any size. They work best with the SVG format, but can also export to bitmap formats when needed. The recommended workflow is to create your graphics in SVG and export them to PDF before adding them to your LaTeX document:

- **Inkscape** (`https://inkscape.org`) is a **vector graphics editor** for creating scalable artwork such as diagrams, icons, technical drawings, and illustrations. It provides tools for drawing shapes, paths, and text, as well as layers, gradients, and alignment options—features comparable to those found in professional design software.
- **SVG-edit** (`https://github.com/SVG-Edit/svgedit`) is a lightweight, browser-based **SVG editor** with basic features, ideal for quick edits without installation. You can use it at `https://svgedit.netlify.app`.
- **draw.io** (`https://drawio.com`) is a **web-based tool** for creating diagrams, flowcharts, and other structured drawings. You can use it directly in your browser at `https://app.diagrams.net` or install the desktop version, called draw.io Desktop, for offline use.

You can even create vector graphics directly in LaTeX. Since this is a more advanced topic, I wrote an entire book about it: *LaTeX Graphics with TikZ*, released by Packt Publishing. For now, we'll focus on images created outside of LaTeX, such as photos and illustrations.

Embedding images

To include images, the standard package to use is graphicx. The x in the name indicates that it extends the original (now outdated) graphics package.

Let's walk through a short document where we insert a picture between two paragraphs. Follow these steps:

1. Start a new document and load the babel and blindtext packages to print some placeholder text, as follows:

   ```
   \documentclass[a5paper]{article}
   \usepackage[english]{babel}
   \usepackage{blindtext}
   \usepackage{graphicx}
   \pagestyle{empty}
   \begin{document}
   \section{Including a picture}
   \blindtext
   ```

2. Begin a figure environment and use \centering to center the image:

   ```
   \begin{figure}
     \centering
   ```

3. Insert the image using the \includegraphics command with the filename as the argument. We will use example-image as the filename since that's a sample image included with TeX Live:

   ```
     \includegraphics[width=4cm]{example-image}
   ```

4. Add a caption, end the figure environment, and end the document with filler text:

   ```
     \caption{Test figure}
   \end{figure}
   \blindtext
   \end{document}
   ```

5. Click **Typeset** to compile the document and check the result, as shown in the following screenshot:

Figure 1: Test figure

1 Including a picture

Hello, here is some text without a meaning. This text should show what a printed text will look like at this place. If you read this text, you will get no information. Really? Is there no information? Is there a difference between this text and some nonsense like "Huardest gefburn"? Kjift – not at all! A blind text like this gives you information about the selected font, how the letters are written and an impression of the look. This text should contain all letters of the alphabet and it should be written in of the original language. There is no need for special content, but the length of words should match the language. Hello, here is some text without a meaning. This text should show what a printed text will look like at this place. If you read this text, you will get no information. Really? Is there no information? Is there a difference between this text and some nonsense like "Huardest gefburn"? Kjift – not at all! A blind text like this gives you information about the selected font, how the letters

Figure 6.1 – An image in a document

The key command here is `\includegraphics`, where we specified the name of the file to be included. LaTeX loads this file if it exists; otherwise, it will display an error message. LaTeX supports the following file types we mentioned before:

- PNG, JPG, and PDF if you directly compile to PDF (using pdflatex)
- EPS if you compile to DVI and convert to PS and PDF (traditional LaTeX)

You don't need to specify a filename extension, as LaTeX automatically searches for supported extensions. If multiple files share the same name but have different extensions, pdflatex selects the first one it finds in this order: `.pdf`, `.png`, `.jpg`, `.jpeg`.

Either put the file into the same directory as your document or specify a full or relative pathname, as follows:

```
\includegraphics{appendix/figure1}
```

In file paths, use forward slash characters (/); don't use backslash characters (\), as the latter begins a LaTeX command.

Go ahead and copy a picture of your choice into your document directory. Give `\includegraphics` its filename, and compile. LaTeX embeds the picture with its original size.

In the following sections, we will see how to add an image of a specified size, include a whole PDF page, or place it in the background behind the text.

Scaling an image

When inserting an image, you may choose a different size. The `\includegraphics` command supports this with a list of key-value options, as follows:

```
\includegraphics[key=value list]{file name}
```

The `graphicx` package documentation, available at `http://texdoc.org/pkg/graphicx`, lists all keys and possible values. Here are the most popular ones and what they do:

- `width`: Image width, for example, `width=0.9\textwidth`
- `height`: Image height, for example, `height=3cm`
- `scale`: Scaling factor, for example, `scale=0.5`
- `angle`: Rotation angle, for example, `angle=90`

There are additional options, such as for clipping, but you can easily do such processing with any graphics software before you include it.

Instead of turning a figure by 90 degrees, you could also use the `sidewaysfigure` environment of the `rotating` package (see `https://texdoc.org/pkg/rotating`).

Including entire pages

What if we want to include pictures that are wider or taller than the text area? While the `\includegraphics` command can handle this, LaTeX may complain about the width or size and might even move them to the next page if there's not sufficient space.

Using the `pdfpages` package, we can insert large images and even entire pages. The `pdfpages` package provides the `\includepdf` command, which can embed a single page, a partial page, or an entire multi-page PDF document at once. Despite its name, it can also include PNG and JPG files, not just PDF files.

An example of its basic usage could look like this:

```
\usepackage{pdfpages}
...
\includepdf[pages=-]{contract}% include entire contract.pdf
\includepdf[pages=2-4]{spec}% include pages 2-4 of spec.pdf
```

A common use is combining several PDF files into a single PDF file. We can also use the `pdfpages` package to resize several PDF pages and arrange them on a single sheet. For more details, take a look at the documentation at `https://texdoc.org/pkg/pdfpages`.

Putting images behind the text

Do you need watermarks? Background images? Textboxes positioned at arbitrary positions on the page? And all of this not interfering with other text? The `eso-pic` package does this for you.

In *LaTeX Cookbook*, you can read a step-by-step example of how to do this in the *Absolute positioning of text* section in *Chapter 2, Tuning the Text*.

The `textpos` package offers another approach. It's developed for placing boxes with text or graphics at absolute positions on a page; see `https://texdoc.org/pkg/textpos`.

With a modern LaTeX installation, you can use a **hook**. That's a place in the code where other code can be inserted to run at that moment, such as a **callback** or **insertion point**. In short, this is like `\AddToHook{shipout/background}{<your image commands>}`.

Now, we will take a look at dynamic image positioning.

Managing floating figures

When a page break occurs, LaTeX can split regular text and continue on the next page. However, it cannot break an image across pages. That's why LaTeX made the `figure` environment that we used in our first example, a **floating environment**. These floating environments are also called **floats** in short. LaTeX may push their content, including their captions, to a place that works well with page layout and page breaks.

Let's look at how to address this.

The `figure` environment takes an optional argument suggesting the final placement of the figure. We will test the effect in our earlier example, as shown here:

1. Return to the previous example from *Figure 6.1*. This time, add the `h` and `t` options in the highlighted line (where `h` and `t` stand for *here* and *top*), as follows:

```
\begin{figure}[ht]
  \centering
  \includegraphics{example-image}[width=4cm]
  \caption{Test figure}
\end{figure}
```

2. Compile the document and take a look at the output. It should look like this:

1 Including a picture

Hello, here is some text without a meaning. This text should show what a printed text will look like at this place. If you read this text, you will get no information. Really? Is there no information? Is there a difference between this text and some nonsense like "Huardest gefburn"? Kjift – not at all! A blind text like this gives you information about the selected font, how the letters are written and an impression of the look. This text should contain all letters of the alphabet and it should be written in of the original language. There is no need for special content, but the length of words should match the language. Hello, here

Figure 1: Test figure

is some text without a meaning. This text should show what a printed text will look like at this place. If you read this text, you will get no information. Really? Is there no information? Is there a difference between this text and some nonsense like "Huardest gefburn"? Kjift – not at all! A blind text like this gives you information about the selected font, how the letters

Figure 6.2 – An image within the text

3. Change the options in the highlighted line to `!b` (where `b` stands for *bottom*), as follows:

```
\begin{figure}[!b]
```

4. Compile again. The figure is now forced to float to the bottom, and now we see this page:

1 Including a picture

Hello, here is some text without a meaning. This text should show what a printed text will look like at this place. If you read this text, you will get no information. Really? Is there no information? Is there a difference between this text and some nonsense like "Huardest gefburn"? Kjift – not at all! A blind text like this gives you information about the selected font, how the letters are written and an impression of the look. This text should contain all letters of the alphabet and it should be written in of the original language. There is no need for special content, but the length of words should match the language. Hello, here is some text without a meaning. This text should show what a printed text will look like at this place. If you read this text, you will get no information. Really? Is there no information? Is there a difference between this text and some nonsense like "Huardest gefburn"? Kjift – not at all! A blind text like this gives you information about the selected font, how the letters

Figure 1: Test figure

Figure 6.3 – An image at the bottom of the page

By adding some characters standing for placement options, we could force the figure to appear wherever we want.

There is a starred form—namely, `figure*`—in a two-column layout that puts a figure into a single column. In one-column mode, there's no difference from the non-starred form.

Let's now take a closer look at the positioning of figures. We will see how we can set preferences for where figures appear, such as at the top or bottom of the page, force immediate output, or at least limit floating, and arrange images next to each other or within the text flow.

Understanding placement options

The optional argument of the `figure` environment specifies where LaTeX can position the figure. Four letters stand for four possible places, as outlined here:

- `h` stands for *here*. The figure may appear right where we wrote it in the source code.
- `t` stands for *top*. It permits placing the figure at the top of a page.
- `b` stands for *bottom*. It allows he figure to appear at the bottom of a page.
- `p` stands for *page*. The figure can be placed on a separate page, where only floats may reside, but no regular text.

A fifth option might come in handy:

- `!` overrides LaTeX defaults; it tells LaTeX to relax some internal constraints, increasing the chance of placing the float earlier.

If you don't specify any option, LaTeX might place the figure far away, which can be surprising to new LaTeX users. Specifying more options helps to keep it as near as possible. The most flexible option is to use the `[!htbp]` placement, which allows the figure to go virtually anywhere. You can still consider removing a placement specifier if you don't like the outcome.

Forcing the output of figures

LaTeX holds figures and tables until they fit the page layout without violating LaTeX's typographic rules. If you want to force LaTeX to print them, there's a command to do so: the `\clearpage` command ends the current page and outputs all queued figures and tables. You can also use `\cleardoublepage`, which does the same but in a two-sided layout. It ensures that the next non-float page is a right-hand page. If necessary, it inserts a blank page.

Ending a page immediately, however, might not be the best approach, as it could leave a lot of empty space on the current page. The `afterpage` package offers an elegant solution. This package can delay the execution of `\clearpage` until the current page is ended, finished, as shown:

```
\usepackage{afterpage}
...
body text
\afterpage{\clearpage}
```

We may not need to use the `afterpage` package often, as we can simply run a `\clearpage` command at the desired location, such as at the end of a section. We can automate that case. Let's look at this in the next section.

Limiting floating

As previously said, a figure may float far away, possibly even into the next section. The placeins package helps to restrict the floating. If you load placeins with \usepackage{placeins} and write \FloatBarrier somewhere in your document, no figure can float beyond that. This macro ensures that floats stay closer to their original place.

A very convenient way to prevent floats from crossing section boundaries is by stating the section option, as follows:

```
\usepackage[section]{placeins}
```

This option inserts an implicit \FloatBarrier at the beginning of each section. Two further options—namely, above and below—allow you to lower the restrictions, preventing floats from appearing above the start of the current section or below the beginning of the next section.

Figures don't pass into the next chapter because the \chapter command includes an implicit \clearpage.

Avoiding floating altogether

Do you want an image to appear exactly where you placed it in the code? The obvious answer is: in that case, don't use a floating figure environment. You can use \includegraphics without a figure environment. For example, you could include and center an image by doing the following:

```
\begin{center}
  \includegraphics[width=4cm]{example-image}
\end{center}
```

Captions are designed for floating environments, though, so a \caption command won't work here. If you still want to have a caption, you may use the \captionof command. The caption package, **KOMA-Script** classes, introduced in *Chapter 3, Designing Pages*, and the tiny capt-of package provide that command, which you can use in the following way:

```
\usepackage{capt-of}% or caption
...
\begin{minipage}{\linewidth}
  \centering
  \includegraphics{example-image}
  \captionof{figure}{Test figure}
\end{minipage}
```

The `minipage` environment keeps a picture and its caption together because no page break can occur in a `minipage` environment. The definition of `\captionof` is the same as `\caption`, except that there is an additional argument specifying the float type—in this case, `figure`, as illustrated in the following code snippet:

```
\captionof{figure}[short text]{long text}
```

Be aware that the numbering could become inconsistent if you mix real floats and fixed figures. As you would give up LaTeX's positioning capabilities, you have to take care that pages are still adequately filled.

The `float` package provides a convenient and consistent-looking approach to this. It introduces the `H` placement option, causing the float to appear right there, as illustrated in the following code snippet:

```
\usepackage{float}
...
\begin{figure}[H]
  \centering
  \includegraphics{example-image}
  \caption{Test figure}
\end{figure}
```

Choose the method that best fits your needs. When I first started with LaTeX, I used to think "don't use a floating environment when you don't want it to float" and preferred using `minipage` instead. These days, I recommend a `figure` environment with the `H` option for consistent syntax, and `H` is easy to change later into other options.

For more detailed information about float placement, see `https://latex.net/floats`, which collects explanations in several languages.

Arranging multiple images

To group several subfigures with individual captions inside a single figure, LaTeX offers several supporting packages you can choose from, outlined as follows:

- `subcaption` is a package for subfigures with individual subcaptions and belongs to the `caption` package. If you use hyperlinks in a document, this is the recommended choice, as it best supports hyperlinks. For hyperlinks, also see *Chapter 12, Using Hyperlinks and Designing Headings*.

- `subfig` is a sophisticated package that supports the inclusion of small figures. It handles positioning, labeling, and captioning within a single float.
- `subfigure` is still available for this purpose and you will find it online, but it's no longer maintained, and you can consider it obsolete since `subfig` has appeared.

Don't load two of these packages together. In general, loading two packages that serve the same purpose can lead to conflicts.

For aligning images, stacking images, or positioning them in a grid, examples are provided and explained in *LaTeX Cookbook, Chapter 5, Working with Images.*

Letting text flow around images

If you want to add a playful or visually appealing layout, you can let the text flow around an image or figure. We can achieve this using the `wrapfig` package and its `wrapfigure` environment.

Let's modify our previous example that embedded an image (see *Figure 6.3*). We would like the image to appear on the left side, accompanied by the body text on the right side, with the help of the following steps:

1. In our example code for *Figure 6.3*, additionally load the `wrapfig` package, as follows:

```
\documentclass[a5paper]{article}
\usepackage[english]{babel}
\usepackage{blindtext}
\usepackage{graphicx}
\usepackage{wrapfig}
\pagestyle{empty}
\begin{document}
```

2. Start an unnumbered section, and place a `wrapfig` environment within some placeholder text, like this:

```
\section*{Text flowing around an image}
\blindtext
\begin{wrapfigure}{l}{4cm}
  \includegraphics[width=4cm]{ example-image}
  \caption{Test figure}
\end{wrapfigure}
\blindtext
\end{document}
```

3. Compile the document and take a look. You should see the following output:

Text flowing around an image

Hello, here is some text without a meaning. This text should show what a printed text will look like at this place. If you read this text, you will get no information. Really? Is there no information? Is there a difference between this text and some nonsense like "Huardest gefburn"? Kjift – not at all! A blind text like this gives you information about the selected font, how the letters are written and an impression of the look. This text should contain all letters of the alphabet and it should be written in of the original language. There is no need for special content, but the length of words should match the language.

Figure 1: Test figure

Hello, here is some text without a meaning. This text should show what a printed text will look like at this place. If you read this text, you will get no information. Really? Is there no information? Is there a difference between this text and some nonsense like "Huardest gefburn"? Kjift – not at all! A blind text like this gives you information about the selected font, how the letters are written and an impression of the look. This text should contain all letters of the alphabet and it should be written in of the original language. There is no need for special content, but the length of words should match the language.

Figure 6.4 – Text flowing around an image

The `wrapfigure` environment options differ from the `figure` environment. We used just two of them. If you need more, here's the complete definition:

```
\begin{wrapfigure}[number of lines]{placement}[overhang]{width}
```

The first optional argument states the number of wrapped text lines. If omitted, this would be automatically calculated from the height. The second argument, `placement`, can be one of the `r`, `l`, `i`, or `o` characters for the right, left, inner, or outer side, or the corresponding uppercase letters `R`, `L`, `I`, or `O`, with the same meaning, but allowing the figure to float. Only one character is allowed for specifying the option. The other optional argument, `overhang`, can set a width by which the figure should hang out into the margin; the default is `0` pt. The final—and mandatory—argument gives the width of the figure.

You can read more in the manual at `https://texdoc.org/pkg/wrapfig`.

Summary

In this chapter, we covered how to include images in our LaTeX documents. You learned which image formats are supported and how to control figure positioning.

LaTeX can automatically generate a list of figures, similar to a table of contents. We will deal with such lists in *Chapter 7, Using Cross-References*.

Since figures are numbered automatically, you can cross-reference them within your text. In the next chapter, you'll learn how to do that using LaTeX's built-in referencing features.

7

Using Cross-References

In the previous chapters, we created figures and tables and added captions to them. LaTeX automatically takes care of numbering. Documents can also include many other numbered elements, such as pages, sections, and list items. And there's more: if you're writing mathematical content, you may want to number equations, theorems, definitions, and other elements as well.

Numbering isn't just for counting; it also allows us to refer to these elements elsewhere in the document. For instance, in this chapter, if I wanted to direct you to the third figure, I would write "See *Figure 7.3*." If you insert another figure, LaTeX will update the numbering of all subsequent figures. But what about the references? LaTeX handles all our cross-references and will update them. So, we should not write a plain reference manually like *7.3*, but instead assign internal **labels** that we can **cross-reference** later.

In this chapter, we'll cover the following:

- Setting labels and references
- Using advanced referencing
- Referring to labels in other documents
- Checking labels and references
- Turning references into clickable hyperlinks

We'll work through this using hands-on examples.

Setting labels and references

To reference a specific part of your document, you first need to mark it using a **label**. The name you assign to this label is what you'll use later to reference it.

Let's walk through a practical example. We'll typeset a list of the most commonly used LaTeX packages, based on a survey made on `https://latex.org`. Using the `\label` command, we'll mark positions we want to refer to later using the `\ref` command. Here's how:

1. Begin a new book document:

```
\documentclass{book}
\begin{document}
```

2. Start a chapter and a section, and add a label to this section:

```
\chapter{Statistics}
\section{Most used packages by LaTeX.org users}
\label{sec:packages}
```

3. Add some body text and a footnote:

```
The Top Five packages, used by LaTeX.org
members\footnote{according to the 2021 survey on
LaTeX.org\label{fn:project}}:
```

4. Write an enumerated list and put a label on a few items:

```
\begin{enumerate}
  \item graphicx\label{item:graphicx}
  \item babel
  \item amsmath\label{item:amsmath}
  \item geometry
  \item hyperref
\end{enumerate}
```

5. Add another chapter with a label:

```
\chapter{Mathematics}
\label{maths}
```

6. Finish with some text, and refer back to earlier items using the \ref and \pageref commands:

```
\emph{amsmath}, on position \ref{item:amsmath}
of the top list in section~\ref{sec:packages} on
page~\pageref{sec:packages}, is indispensable to
high-quality mathematical typesetting in \LaTeX.
\emph{graphicx}, on position \ref{item:graphicx},
is for including images. See also the footnote
\ref{fn:project} on page~\pageref{fn:project}.
\end{document}
```

7. Compile the document and examine the output. Page 1 starts with headings and a bullet list.

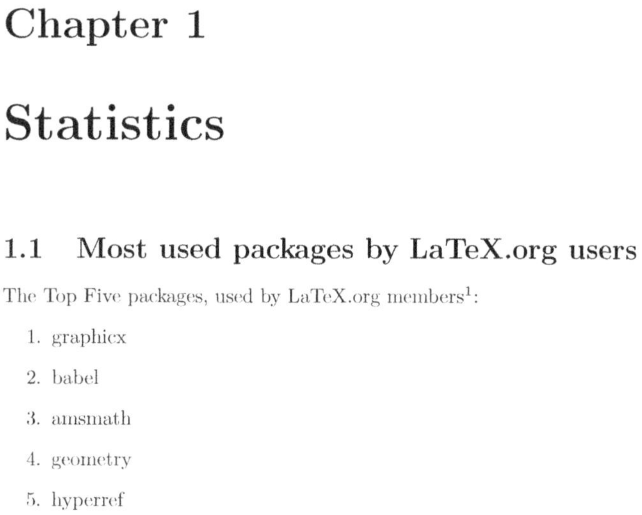

Chapter 1

Statistics

1.1 Most used packages by LaTeX.org users

The Top Five packages, used by LaTeX.org members[1]:

1. graphicx
2. babel
3. amsmath
4. geometry
5. hyperref

Figure 7.1 – First chapter

Page 1 ends with a footnote:

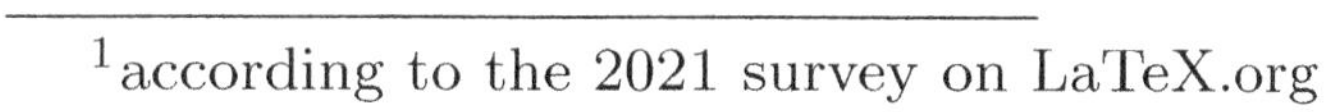

[1]according to the 2021 survey on LaTeX.org

Figure 7.2 – Footnote

Page 2 is empty since *Chapter 2* starts on a right-hand page, which is page 3:

Chapter 2

Mathematics

amsmath, on position ?? of the top list in section ?? on page ??, is indispensable to high-quality mathematical typesetting in LaTeX. *graphicx*, on position ??, is for including images. See also the footnote ?? on page ??.

Figure 7.3 – Unresolved references

8. Do you see the question marks? Here, references are still missing. Compile again and compare:

Chapter 2

Mathematics

amsmath, on position 3 of the top list in section 1.1 on page 1, is indispensable to high-quality mathematical typesetting in LaTeX. *graphicx*, on position 1, is for including images. See also the footnote 1 on page 1.

Figure 7.4 – Resolved references

We created cross-references using these three commands:

- `\label` marks the position
- `\ref` prints the number of the element associated with a label
- `\pageref` prints the page number where that label occurs

Each of these commands takes the label name as an argument. We can define the names freely.

We had to compile twice because LaTeX requires one run to generate the references that it can read during the next compilation run. If LaTeX can't resolve a reference yet, you'll see two question marks in the output.

Let's now have a closer look at how to create anchor labels and refer to them effectively.

Assigning a label

The \label{name} command assigns the current position in the document to the label name. Precisely, it does the following:

- If you use the \label command in regular text, the label refers to the current sectional unit, such as the chapter or section
- If the \label command is used inside a numbered environment, such as a figure, table, or numbered list, the label will refer to that environment or item

So, we cannot label a section within a table environment. To avoid any problems because of possible unsuitable positioning, a good rule of thumb is to place the \label command right after the position that we would like to refer to. For instance, place it directly after the corresponding \chapter or \section command.

In the figure or table environments, the \caption command is implicitly responsible for the numbering. That's why \label has to be placed after \caption, not before.

Floating environments may look like the following:

```
\begin{figure}[htbp!]
\centering
\includegraphics{filename}
\caption{Test figure}\label{fig:name}
\end{figure}
```

Here's how a table with a label may look:

```
\begin{table}[htbp!]
\centering
\caption{table descripion}\label{tab:name}
\begin{tabular}{cc}
...
\end{tabular}
\end{table}
```

Label names may include letters, numbers, and punctuation marks, and they are case-sensitive.

In larger documents, the number of labels could be very high, and label management becomes essential. To avoid naming conflicts, it's a good idea to prefix labels based on the type of environment they represent. It has become common practice to label equations with `eq:name`, figures with `fig:name`, tables with `tab:name`, and sections with `sec:name`, with similar approaches in other cases. Keep in mind that labels such as `fig:1` defeat the purpose, as the idea is to avoid manually setting numbers entirely.

In the following sections, we'll look at several ways to reference labels.

Referring directly to a labeled position

After assigning a label with a name, we can refer to that name. For this, we use the `\ref{name}` command. This command prints the number associated with that label, `name`, which can be, for example, the number of an environment, a list item, or a sectioning element. We can even use `\ref{name}` before the corresponding `\label{name}` command appears in the code.

Although the mechanism is simple, it's quite powerful. Each time we compile a document, LaTeX checks the labels and reassigns the numbers, reflecting all changes in numbering. If LaTeX detects modified labels, it will print a message that another compiler run is required to update the corresponding labels. When in doubt, compile twice.

Referring to a page

The `\pageref{name}` command works similarly to `\ref`, but instead of returning a number for an element, such as an environment or a section, it prints the page number where the label has been set.

Would all the references still be accurate if we changed the section and page numbers? Let's try that. Insert a dummy section and a page break at the beginning of our chapter, highlighted here:

```
\chapter{Statistics}
\section{Introduction}
\newpage
\section{Most used packages by LaTeX.org users}
\label{sec:packages}
```

Click on **Typeset** once. LaTeX will compile it, but it will show a message:

```
LaTeX Warning: Label(s) may have changed. Rerun to get cross-references
right.
```

That's what we shall do! Click on **Typeset** a second time, and now all the numbers have been correctly adjusted to **section 1.2** and **page 2**:

amsmath, on position 3 of the top list in section 1.2 on page 2, is indispensable to high-quality mathematical typesetting in LaTeX. *graphicx*, on position 1, is for including images. See also the footnote 1 on page 2.

Figure 7.5 – Automatically adjusted references

You can combine both commands. If you'd like to refer to both the section and page number in one sentence, you can write the following:

```
See figure~\ref{fig:name} on page~\pageref{fig:name}.
```

As you know how to define a command, you could make such referencing easier:

```
\newcommand{\fullref}[1]{\ref{#1} on page~\pageref{#1}}
...
See figure~\fullref{fig:name}.
```

This gives you a comprehensive reference, such as *See Figure 4.2 on page 32*. However, if the reference, such as for a figure, appears on the same page, writing out the page number looks a bit odd. How can we avoid that? The `varioref` package offers an elegant solution. We will focus on such advanced referencing in the following sections.

Using advanced referencing

LaTeX helps with automating all kinds of references. It's not limited to numbering. LaTeX can even generate names and phrases. We will dig deeper into that here.

Producing smart page references

The `varioref` package offers commands to add *on the preceding page*, *on the following page*, or the page number to a reference, depending on the context.

We will use the two `varioref` commands to introduce variable references, `\vref`, and `\vpageref`, to achieve enhanced reference texts, as follows:

1. Open our current example from the first section. Add the `varioref` package to the preamble. Use the nospace package option, which ensures that `varioref` doesn't insert additional undesirable space before or after a reference:

   ```
   \usepackage[nospace]{varioref}
   ```

2. Modify the content of the second chapter in our example code:

```
\emph{amsmath}, on position \vref{item:amsmath}
of the top list in section~\vref{sec:packages},
is indispensable to high-quality mathematical
typesetting in \LaTeX. \emph{graphicx}, on position
\vref{item:graphicx}, is for including images.
See also the footnote \vref{fn:project}, that is,
\vpageref{fn:project}.
```

3. Compile twice, and look at the result:

amsmath, on position 3 on the facing page of the top list in section 1.2 on the preceding page, is indispensable to high-quality mathematical typesetting in LaTeX. *graphicx*, on position 1 on the facing page, is for including images. See also the footnote 1 on the preceding page, that is, on the facing page.

Figure 7.6 – An image at the bottom of the page

The `\vref` command checked the distance from the label of the referenced section. As the label is on the facing page (here, on the preceding page in a two-sided layout), `\vref` wrote **1.2 on the preceding page**.

`\vpageref` refers to **the facing page** at the end of the paragraph.

`\vref{name}` works as follows:

- If the reference and `\label{name}` are on the same page, it behaves exactly like `\ref` and omits the page number.
- If the reference and the corresponding `\label` are on adjacent pages, `\vref` prints the referred number and, additionally, *on the preceding page, on the following page*, or *on the facing page*. It will choose the latter if the document is two-sided, that is, if `\label` and the reference fall onto a double-page spread.
- If they are farther apart, it will print both `\ref` and `\pageref`.

`\vpageref` is similar to `\pageref` but behaves like `\vref` concerning the page reference.

`varioref` switches between phrasings to have a bit of variation. It can say *following page* or *next page, preceding page* or *previous page, this page* or *current page*. And in a double-page layout, it switches between *facing page* and *preceding page* or *next page*. With such variation, the text reads more naturally. You can see this alternating phrasing in *Figure 7.6*.

Even when using `varioref`, you may still use the standard `\ref` and `\pageref` commands along-side the new commands.

Fine-tuning page references

When a label and its reference are located close to each other, they may fall on the same page, but not always. In such cases, we usually know whether the label appears above or below the reference. The `varioref` package allows indicating this using an optional argument to the `\vpageref` command, as shown below:

```
see the figure \vpageref[above]{fig:name}
```

This will print the following:

- *see the figure above*, if the figure is on the same page
- *see the figure on the page before*, if the figure is on the preceding page

Whereas with the following code, we will have a different output:

```
see the footnote \vpageref[below]{fn:name}
```

This will instead print the following:

- *see the footnote below*, if the footnote is on the same page
- *see the footnote on the following page*, if the footnote is on the next page

The `\vpageref` command understands two optional arguments. While in the first optional argument, we can state a phrase if the label and reference fall on the same page, in the second optional argument, we can give a phrase for the case when the label and reference fall on different pages. So, we could even write the following:

```
see the \vpageref[above figure][figure]{fig:name}
```

This would print the following:

- *see the above figure*, if the figure is on the same page
- *see the figure on the previous page*, if the figure is on the preceding page

All of this may seem a bit elaborate, but your demands might increase over time, requiring more sophisticated referencing so that these features might come in handy someday.

Referring to page ranges

`varioref` offers two more commands:

- `\vpagerefrange[opt]{label1}{label2}`, where `label1` and `label2` denote a range (such as a sequence of figures from `fig:a` to `fig:c`). If both labels fall on the same page, the result is the same as with `\vpageref`. Otherwise, the output will be a range, such as *on pages 32-36*. `opt` would be used if both labels fall on the current page.
- `\vrefrange[opt]{label1}{label2}` is analogous to `\vref`: `see figures \vrefrange{fig:a}{fig:c}` may result in *see figures 4.2 to 4.4 on pages 36-37*.

Visit `https://latexguide.org/chapter-07` to see examples.

You can find more information regarding customization in the package manual. As usual, you can open it at the command prompt by typing `texdoc varioref` or by visiting `https://texdoc.org/pkg/varioref`.

Using automatic reference names

Tired of repeatedly writing `figure~\ref{fig:name}` and `table~\ref{tab:name}`? Wouldn't it be great if LaTeX could automatically determine the element type meant by `\ref{name}` and automatically insert the correct reference text with the element name and number for you? What if we want to abbreviate, say, `fig.~\ref{fig:name}` in the whole document? The `cleverev` package makes this easy. It automatically detects the type of cross-reference and the context in which it is used.

You can use `\cref` or `\Cref` instead of `\ref`; choose the latter if you wish to capitalize. The corresponding range commands are `\crefrange` and `\Crefrange`.

Let's rewrite our first example to refer using `cleveref`. To verify that the package acts well, we intentionally omit prefixes in the label names for `\label` and `\cref`, as shown here:

1. Modify our first example of this chapter in the following way. The changed lines are highlighted:

   ```
   \documentclass{book}
   \usepackage{cleveref}
   \crefname{enumi}{position}{positions}
   \begin{document}
   \chapter{Statistics}
   \label{stats}
   \section{Most used packages by LaTeX.org users}
   ```

```
\label{packages}
The Top Five packages, used by LaTeX.org
members\footnote{according to the 2021 survey on
LaTeX.org\label{project}}:
\begin{enumerate}
  \item graphicx\label{graphicx}
  \item babel
  \item amsmath\label{amsmath}
  \item geometry
  \item hyperref
\end{enumerate}
\chapter{Mathematics}
\label{maths}
\emph{amsmath}, on \cref{amsmath} of the top list in
\cref{packages} of \cref{stats}, is indispensable to
high-quality mathematical typesetting in \LaTeX.
\emph{graphicx}, on \cref{graphicx}, is for
including images.
See also the \cref{project} on \cpageref{project}.
\end{document}
```

2. Click on **Typeset** twice, and check that references have the correct names:

   amsmath, on position 3 of the top list in section 1.1 of chapter 1, is indispensable to high-quality mathematical typesetting in LaTeX. *graphicx*, on position 1, is for including images. See also the footnote 1 on page 1.

Figure 7.7 – Automated references

As we can see, we don't need to specify which object we refer to. `\cref` always chooses the right name and the correct number for us. That's really useful.

We used the `\crefname` command to tell `cleveref` which name it should use for enumerated items. The definition of `\crefname` is as follows:

```
\crefname{type}{singular}{plural}
```

`type` may be one of `chapter`, `section`, `figure`, `table`, `enumi`, `equation`, `theorem`, or many other types we have not encountered yet. `cleveref` uses the singular form for individual references and the plural version for multiple ones. If you need capitalized versions, use `\Crefname`. So, a typical use may be the following:

```
\crefname{figure}{fig.}{figs.}
\Crefname{figure}{Fig.}{Figs.}
```

Also, here, we can refer to ranges using these commands:

- `\crefrange{label1}{label2}` to refer to a range of references
- `\cpagerefrange{label1}{label2}` to refer to a page range

It will get clearer with an example. Let's add this line to our current example:

```
See \crefrange{graphicx}{amsmath} and \cpagerefrange{stats}{maths}.
```

This gives us: *See positions 1 to 3 and pages 1 to 3.*

We can sum up the benefits as follows:

- It reduces the amount of typing.
- We could use arbitrary label names. The `fancyref` package does a similar job but relies on prefixes such as `chap`, `fig`, or `tab`.
- If we decide to change the reference wording, we only need to do it once in the preamble, and it will apply consistently throughout the entire document.

That said, it's still a good idea to use a prefix such as `fig:` or `sec:` to indicate the kind of referenced element. This way, your code would become more understandable – that's why it's considered best practice.

Combining smart references with automatic naming

As the `cleveref` package is fully compatible with `varioref`, you can use both to take full advantage of their features. `cleveref` redefines the `varioref` commands to internally use `\cref`. So, you can use the advanced page referencing features of `varioref` together with the clever naming automation.

Just load `varioref` before `cleveref`, as shown here:

```
\usepackage{varioref}
\usepackage{cleveref}
```

Now, you may use `\vref`, `\cref`, `\ref`, or the other commands – whichever suits your needs best.

Referencing section names and captions

So far, our references have shown only numbers. If you'd like to display the names instead, you can use the `nameref` package. Start by loading it:

```
\usepackage{nameref}
```

Then, use the `\nameref` command in place of `\ref` or `\pageref`, or alongside these commands. In the first example of this chapter, you can write this:

```
See section~\ref{sec:packages}, \nameref{sec:packages},
on page~\pageref{sec:packages}.
```

This gives you: **See section 1.1, Most used packages by LaTeX.org users, on page 1.**

The `\nameref` command works with regular sectioning commands, including `\part` and `\chapter`, as well as with figure and table captions, description list items, and theorem names.

We can also use references to pages, sections, and more across different documents. Let's take a closer look at this in the next section.

Referring to labels in other documents

When working on multiple related documents that refer to each other, you might want to use references to labels of another document. The package named `xr` (short for **external references**) implements this. First, load the `xr` package:

```
\usepackage{xr}
```

If you need to refer to sections or environments in an external document, say, `doc.tex`, insert this command into your preamble:

```
\externaldocument{doc}
```

This enables you to additionally refer to anything that has been given a label in `doc.tex`. You may do this for several documents. If you need to avoid conflicts when an external document uses the same `\label` as the main document, declare a prefix using the optional argument of `\externaldocument`, which you can use to add a prefix. For example, we can use `D-` as a prefix:

```
\externaldocument[D-]{doc}
```

This way, all references from `doc.tex` would be prefixed by `D-`, and you could write `\ref{D-name}` to refer to `name` in `doc.tex`. Instead of `D-`, you may choose any prefix that makes your labels unique.

Checking labels and references

Over time, your document may grow, making it hard to remember all label names and where you referred to them. Fortunately, some tools make it easy to inspect and track them.

Displaying labels

The `showlabels` package displays the labels you defined in your PDF output so that you can see them alongside your text. Just load it:

```
\usepackage{showlabels}
```

When you compile the same document that we used for *Figure 7.1*, you'll see the label names for sections, items, and footnotes right in the output:

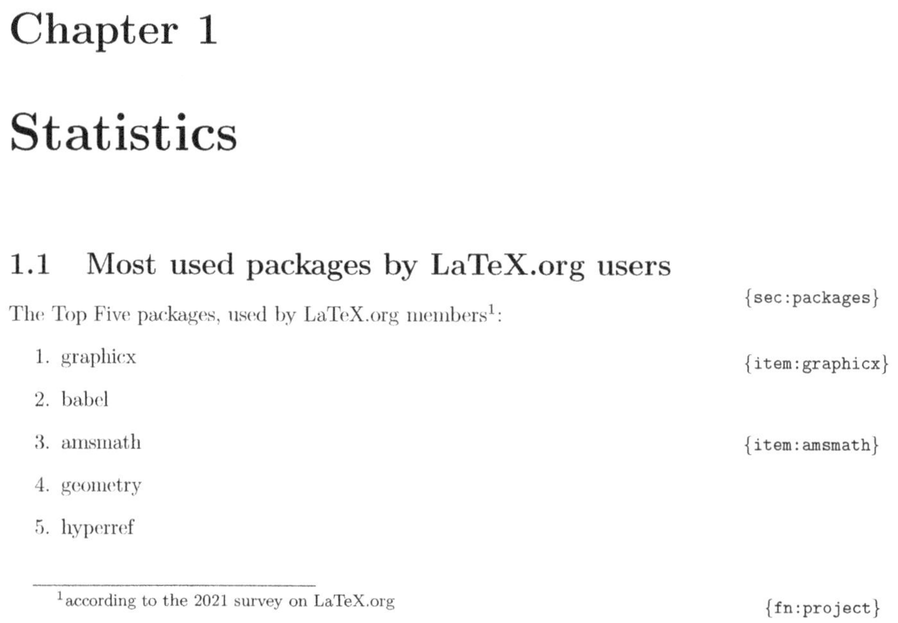

Chapter 1

Statistics

1.1 Most used packages by LaTeX.org users {sec:packages}

The Top Five packages, used by LaTeX.org members[1]:

1. graphicx {item:graphicx}
2. babel
3. amsmath {item:amsmath}
4. geometry
5. hyperref

[1] according to the 2021 survey on LaTeX.org {fn:project}

Figure 7.8 – Displayed labels

This can be very useful while writing a longer document. Just be sure you remove it or comment it out before you compile and deliver your final version.

You can customize the appearance of the labels, like this, for example, to see them in bold red:

```
\renewcommand{\showlabelfont}{\small\bfseries\color{red}}
```

Another option is to use the `showkeys` package:

```
\usepackage{showkeys}
```

It also shows the labels in the margin, where you defined them, but additionally, it displays the names where you reference them, like this:

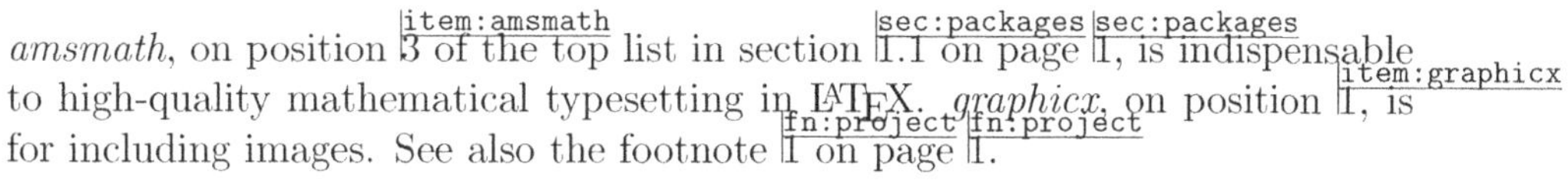

item:amsmath sec:packages sec:packages
amsmath, on position 3 of the top list in section 1.1 on page 1, is indispensable
item:graphicx
to high-quality mathematical typesetting in LaTeX. *graphicx*, on position 1, is
fn:project fn:project
for including images. See also the footnote 1 on page 1.

Figure 7.9 – Displayed cross-references

This can also help you develop your document.

Identifying duplicate labels

Labels are meant to be unique identifiers so that LaTeX can reliably decide which object a `\ref` or `\pageref` should point to. If you use the same label name more than once, LaTeX issues a warning such as `Label name multiply defined`. The consequence of such duplicate labels is that cross-references may silently point to the wrong section, figure, table, or equation. LaTeX does not stop compiling, but the resulting output can be inconsistent or misleading.

The warning is written to the `.log` file, and any serious LaTeX editor will surface duplicate labels as warnings, making them easy to spot and fix.

Prefixing labels with type (`sec:`, `fig:`, `tab:`, `eq:`), as mentioned earlier in this chapter, dramatically reduces collisions in larger documents.

Verifying references

To check whether all labels and references in your document are actually used, you can load the refcheck package. It helps you identify unused labels, unreferenced equations, and missing references. Let's say you added `\label{item:hyperref}`, but you never referred to it.

Simply load the package:

```
\usepackage{refcheck}
```

You will see the names of the used labels, and any unused labels will appear with question marks, making them easy to spot:

1.1 Most used packages by LaTeX.org users

⟨sec:packages⟩ The Top Five packages, used by LaTeX.org members[1]:

⟨item:graphicx⟩
1. graphicx
2. babel

⟨item:amsmath⟩
3. amsmath
4. geometry

?⟨item:hyperref⟩?
5. hyperref

Figure 7.10 – Highlighting labels without reference

In the next section, we'll see how to make references clickable so readers can jump directly to the referenced element.

Turning references into hyperlinks

PDF can include clickable links and bookmarks. How about exploring that? There's an outstanding package that offers hyperlink support – the `hyperref` package.

Try it by loading `hyperref` right before `cleveref`. This order is essential for the references to work because `cleveref` detects whether `hyperref` has been loaded and makes the references to hyperlinks. Even without any options or commands, your document will be hyperlinked as much as possible because of the following:

- All references become hyperlinks. Click on any of those numbers to jump to the referred table, list item, section, or page.
- Each footnote marker is a hyperlink to the footnote text. Click it to jump there.
- If you insert `\tableofcontents`, you will get a bookmark list for the documents, chapters, and sections listed in a navigation bar of your PDF reader.

`hyperref` can do even more for you – linking index entries to text passages, back-referencing of bibliography entries, and more. You can finely customize the behavior using options such as, for instance, choosing the color or frames for hyperlinks. So, you should keep that valuable package in mind. In *Chapter 12*, *Using Hyperlinks and Designing Headings*, we shall return to this topic.

On the one hand, hyperref detects many other packages, such as varioref, and can turn their commands into hyperlinks; that's why we should load hyperref after most other packages. On the other hand, some exceptional packages, such as cleveref, detect hyperref functions and build on them – in such cases, we should load them after hyperref. So, if you combine varioref, cleveref, and hyperref, the package loading order should be as follows:

```
\usepackage[nospace]{varioref}
\usepackage{hyperref}
\usepackage{cleveref}
```

The hyperref package manual has a whole section about compatibility with other packages and the required loading order. Open it by entering texdoc hyperref at the command prompt or visit https://texdoc.org/pkg/hyperref. Most of the time, hyperref should be loaded last, with a few exceptions mentioned in the manual.

Summary

In this chapter, we discussed how to reference chapters, sections, footnotes, and environments by their number or by the number of the corresponding page.

By using labels and names for referencing, we did not need to specify a number ourselves; LaTeX determines the correct page number, section number, footnote number, or environment number for us.

You also learned some innovative ways to handle context-aware referencing.

In the next chapter, we'll move on to working with lists, which consist mainly of references: tables of contents, lists of figures and tables, and bibliographies.

8

Managing Contents, Indexes, and Bibliography

So far, you have learned how to structure and format documents, create tables and images, and connect their parts using cross-references. This chapter shows how LaTeX builds on that by automatically collecting and presenting a document's content in tables of contents, figures, tables, indexes, and bibliographies. LaTeX makes it easy to create all kinds of cross-referenced lists. For example, we've already seen that using just the `\tableofcontents` command gives you a clean, nicely formatted **table of contents** (**TOC**). It automatically collects the chapter and section headings, along with their page numbers, and generates a tidy list.

A TOC and an **index** help readers navigate a document. A **list of tables** and a **list of figures** work in the same way. Academic papers and books often include a list of other works to cite their sources or support the topic. If this list contains only cited works, we often call it a **list of references**; if it also contains uncited but relevant works, we call it a **bibliography**.

By the end of this chapter, you'll know how to create and customize all of these lists.

Here's what we'll cover:

- Customizing the table of contents
- Generating an index
- Creating a bibliography
- Adding a glossary
- Changing the headings

Let's begin with the contents.

Customizing the table of contents

Simply typing the `\tableofcontents` command produces a ready-made list of contents. But LaTeX also lets us fine-tune its appearance and structure. Let's see how.

We'll create a short example document that serves as our test file throughout this chapter.

It will contain a few headings so that we can experiment with the table of contents and learn how to adjust it. We will modify it to be more nuanced and include additional entries.

In *Chapter 3*, *Designing Pages*, we saw what happens when we use the `\tableofcontents` command. LaTeX collects all the entries from the headings and automatically lists them. This time, we'll include headings down to the subsubsection level and add a few manual entries. Let's begin by building the base document:

1. Start a new book document:

   ```
   \documentclass{book}
   ```

2. Set the table of contents' depth value to 3 to include headings down to the subsubsection level:

   ```
   \setcounter{tocdepth}{3}
   ```

3. Begin the document:

   ```
   \begin{document}
   ```

4. Print the table of contents at the beginning:

   ```
   \tableofcontents
   ```

5. Write headings in all the levels you want. Use `\addcontentsline` or `\addtocontents` to add something to the TOC manually:

   ```
   \part{First Part}
   \chapter*{Preface}
   \addcontentsline{toc}{chapter}{Preface}
   \chapter{First main chapter}
   \section{A section}
   \section{Another section}
   \subsection{A smaller section}
   ```

```
\subsubsection[Deeper level]{This section has an
even deeper level}
\chapter{Second main chapter}
\part{Second part}
\chapter{Third main chapter}
```

6. Finish with an appendix that has its own chapters:

```
\appendix
\cleardoublepage
\addtocontents{toc}{\bigskip}
\addcontentsline{toc}{part}{Appendix}
\chapter{Glossary}
\chapter{Symbols}
\end{document}
```

7. Click **Typeset** to compile. The first page will show **Contents**, but no entries will appear.
8. Click **Typeset** a second time. Now we can see the table of contents:

Contents

Figure 8.1 – An example of a table of contents

We structured a document using several sectioning commands. LaTeX reads all of our sectioning commands during the first run and creates a `.toc` file that stores all the headings. During the first run, that file didn't exist yet; thus, the TOC remained empty. The second run reads this file and prints the final table of contents.

In our example, we increased the TOC depth by 1 level. We added a chapter-like entry for the preface and inserted a part-like heading for the appendix beginning, using the `\addcontentsline` command. Through the `\addtocontents` command, we inserted some space before the heading. In the following sections, we'll look more closely at these commands and learn how to customize the TOC.

Adjusting the depth of the TOC

Each sectioning command in LaTeX has an associated **TOC level**. These levels determine which headings appear in the table of contents. Here are the values for the standard sectioning commands:

- `\part`: -1 in the `book` and `report` classes, and 0 in the article `class` article
- `\chapter`: 0 (except in the `article` class, since there are no chapters with `article`)
- `\section`: 1
- `\subsection`: 2
- `\subsubsection`: 3
- `\paragraph`: 4
- `\subparagraph`: 5

In the `book` and `report` classes, LaTeX creates TOC entries up to level 2 (i.e., down to the `\subsection` level). In the `article` class, LaTeX creates TOC entries until level 3 by default, that is, down to the `\subsubsection` level. In a book, this means, for example, that `\subsubsection` doesn't generate a TOC entry. There is a variable representing the level, namely, `\tocdepth`. It's an integer variable, which we call a **counter**. To tell LaTeX to include subsubsections in the TOC, we would have to raise this counter. There are two basic ways to adjust a counter value:

- `\setcounter{name}{n}` specifies an integer value of `n` for the `name` counter.
- `\addtocounter{name}{n}` adds the integer value of `n` to the value of the `name` counter. To decrease a counter, choose a negative value for `n`.

For example, the following command makes LaTeX include even `\subparagraph` headings in the TOC:

```
\setcounter{tocdepth}{5}
```

Alternatively, use the `\addtocounter` command to raise or lower the level without needing to know the current value.

In contrast to commands, counter names don't begin with a backslash.

Shortening TOC entries

As you have already learned in *Chapter 3, Designing Pages*, you may choose a text for the TOC different from the heading in the body text. Each sectioning command understands an optional argument for the TOC entry, which is especially useful if you wish to use a very long heading but want a shorter TOC entry. In our example, we did this by means of the following command:

```
\subsubsection[Deeper level]{This section has an even deeper
level}
```

The body text shows the long heading, while the TOC shows the short one. Titles printed at the top of pages, called **running headers**, would also use the short entry. This way, you can avoid a cluttered TOC or overflown page headers.

Adding TOC entries manually

Starred sectioning commands, such as `\chapter*` and `\section*`, are used for printing a heading without generating a TOC entry. In our example, we added one manually using this command:

```
\addcontentsline{file extension}{sectional unit}{text}
```

You can use this command in different situations. The `file extension` specifies which list to update, for example:

- `toc` for the table of contents
- `lof` for the list of figures
- `lot` for the list of tables

The `sectional unit` determines the entry's formatting. For instance, if you set it to `chapter`, LaTeX will format the entry as a regular chapter entry. The same applies to other sectional units, such as `part`, `section`, and `subsection`. The third argument contains the entry text.

You can also insert text or commands more directly using the following command:

```
\addtocontents{file extension}{entry}
```

Unlike `\addcontentsline`, this command writes the argument entry directly into the file as it is, without applying any additional formatting. You can freely define any formatting you like.

You can also use the `\addtocontents` command for various customizations. Here are some examples:

- `\addtocontents{toc}{\protect\enlargethispage{\baselineskip}}` increases the text height so that one more line fits on the contents page.
- `\addtocontents{toc}{\protect\newpage}` forces a page break in the TOC. This can be useful if LaTeX automatically breaks the page in an awkward place, for example, right after a chapter entry, but before the following section entries.
- `\addtocontents{toc}{\protect\thispagestyle{fancy}}` changes the page style of the current TOC page to `fancy`. As the first page of a chapter is of the `plain` style by default, the first page of the TOC would be plain as well, even if you specified `\pagestyle{fancy}` earlier. That command overrides it.

Place such commands where you want their effects to take place. For example, to affect the first TOC page, place it at the beginning of your document. To insert a page break before a specific chapter, put it right before the corresponding `\chapter` command.

Creating and customizing lists of figures

There are two commands for creating lists of figures and tables: `\listoffigures` and `\listoftables`. Depending on the class, they produce a well-formatted list of all captions, along with the figure or table number and the corresponding page numbers. Just like with the TOC, LaTeX handles this automatically. Still, we may use the same techniques, such as with the TOC, to customize the other lists. Let's try that.

Suppose all our figures are diagrams. We'll replace the term *figure* with *diagram*, and we'll add a list of diagrams:

1. Open our current example. Add these lines to your preamble:

   ```
   \renewcommand{\figurename}{Diagram}
   \renewcommand{\listfigurename}{List of Diagrams}
   ```

2. Right after `\tableofcontents`, add the following:

   ```
   \listoffigures
   ```

3. Add a diagram somewhere in the first chapter:

```
\begin{figure}
\centering
\fbox{Diagram placeholder}
\caption{Enterprize Organizational Chart}
\end{figure}
```

4. In the second part of the third chapter, we'd like to add network design diagrams. Let's mark that in the **list of figures (LOF)** and then include the diagrams:

```
\addtocontents{lof}{Network Diagrams:}
\begin{figure}
\centering
\fbox{Diagram placeholder}
\caption{Network overview}
\end{figure}
\begin{figure}
\centering
\fbox{Diagram placeholder}
\caption{WLAN Design}
\end{figure}
```

5. Click **Typeset** twice to get the document and the list:

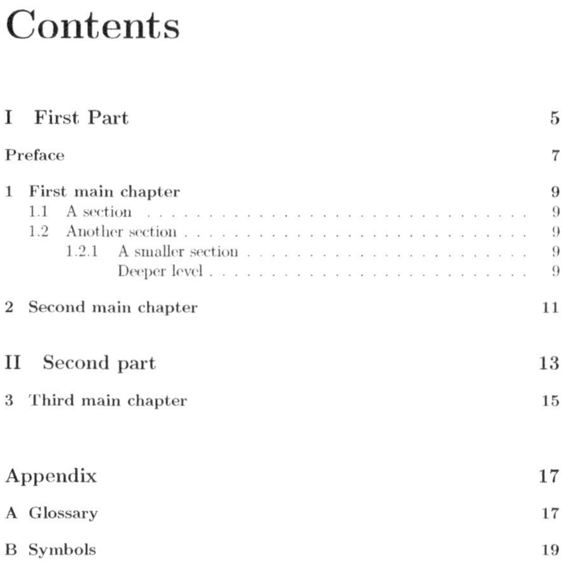
Contents

I First Part 5
Preface 7
1 First main chapter 9
1.1 A section 9
1.2 Another section 9
1.2.1 A smaller section 9
Deeper level 9
2 Second main chapter 11
II Second part 13
3 Third main chapter 15
Appendix 17
A Glossary 17
B Symbols 19

Figure 8.2 – A list of diagrams

We renamed the figures and the list heading by redefining LaTeX macros. At the end of this chapter, you will get a list of names used by LaTeX classes that you may redefine.

As with the TOC, we used the `\addtocontents` command to insert a bold heading into the `.lof` file, where LaTeX stores the figure captions. It works the same way as it does for the TOC.

Creating a list of tables

Creating and customizing a **list of tables** (**LOT**) works just like with figures. The file where LaTeX collects the tables' captions has the `.lot` extension. To modify it, use `lot` as the first argument to the `\addtocontents` command. Everything follows the same logic as before, just like with `\listoftables`, `\tablename`, and `\listtablename`.

Using packages for customization

In addition to the basic methods we've seen, various packages provide sophisticated features for customizing the TOC and the lists of figures and tables:

- `tocloft` gives extensive control over the typography of **TOC**, **LOF**, and **LOT**. You can even define new kinds of such lists.
- `titletoc` provides convenient entry management and is the companion to `titlesec`, an excellent package for customizing sectioning headings.
- `multitoc` offers a layout in two or more columns using the `multicol` package.
- `minitoc` creates small TOCs for each part, chapter, or section.
- `tocbibind` can automatically add a bibliography, index, TOC, LOF, and LOT to the table of contents. It can even number these headings and entries if you prefer.

To learn more, use the `texdoc` command-line tool or visit `https://texdoc.org` to read the package documentation.

Now you know how to create lists of contents, tables, and figures that we usually put at the beginning of a document. Let's continue with lists that appear at the end of a document: a keyword index and a bibliography.

Generating an index

Longer documents often include an **index**. That's a list of words or phrases along with page numbers showing where related topics appear in the document. Unlike a full-text search feature, an index provides carefully chosen pointers to the most relevant information.

Once we identify and mark the words we want to include in the index, LaTeX handles collecting them and typesetting the index.

Let's say our example document contains information about an enterprise, its structure, and its network design. We'll mark places in the text where these concepts occur. Finally, we will tell LaTeX to generate the index. Follow these steps:

1. Go back to our example. In the preamble, load the `index` package and add the command to create the index:

   ```
   \usepackage{index}
   \makeindex
   ```

2. In the caption of our enterprise diagram, index this point with the keyword `enterprise`:

   ```
   \caption{\index{enterprise}Enterprise Organizational
   Chart}
   ```

3. In the third chapter, which contains our diagrams, index by the keyword `network`:

   ```
   \index{network}
   ```

4. Directly before `\end{document}`, create an entry for the index to appear in the table of contents. To ensure that it shows the correct page number, end the page right before it:

   ```
   \clearpage
   \addcontentsline{toc}{chapter}{Index}
   ```

5. On the following line, tell LaTeX to typeset the index:

   ```
   \printindex
   ```

6. If you're using TeXworks, choose **MakeIndex** instead of **pdfLaTeX** in the drop-down box next to the **Typeset** button. Then, click the **Typeset** button. If you use another editor, use its MakeIndex feature or type the following at the command prompt in the document directory, without a file name extension:

   ```
   makeindex documentname
   ```

7. Switch back to **pdfLaTeX**. Click on **Typeset**, and look at the last page:

Index

enterprise, 9

network, 15

Figure 8.3 – An index

We loaded the `index` package, which enhances LaTeX's built-in indexing functions.

Alternatively, you can use the `makeidx` package, which is part of standard LaTeX. The `\makeindex` command prepares the index. Both commands belong to the preamble, so they must be placed before `\begin{document}`.

The `\index` command takes just one argument, namely, the word or the phrase to be indexed. It writes this phrase into a file with the `.idx` extension. If you look into this file, you will find lines such as the following:

```
\indexentry {enterprise}{9}
\indexentry {network}{15}
```

These represent the index entries and their corresponding page numbers.

The external `makeindex` program processes that `.idx` file and generates a `.ind` file. The latter consists of LaTeX code for creating the index. Specifically, it contains the index list environment together with the items and appears as follows:

```
\begin{theindex}
\item enterprise, 9
\indexspace
\item network, 15
\end{theindex}
```

More complex indexes may contain subitems, page ranges, and references to other items. Let's see how to produce such an index. On the book's website, `https://latexguide.org/chapter-08`, you can find fully compilable code containing example commands that we will learn about in the following sections. You can try them out directly on the web page.

Defining index entries and subentries

So far, we've already created simple index entries using the following command:

```
\index{phrase}
```

To produce subentries, specify the main term followed by an exclamation mark, and then the subentry. Here's an example:

```
\index{network!overview}
```

Also, subentries may have subentries; use another exclamation mark, for example:

```
\index{enterprise!organization}
\index{enterprise!organization!sales}
\index{enterprise!organization!controlling}
\index{enterprise!organization!operation}
```

LaTeX supports up to three levels of subentries.

Specifying page ranges

If several pages deal with the same topic, you may specify a page range for the index entry. Use |(where the range starts, and add |) where it ends. So, at the beginning of the `network` chapter, add |(as follows:

```
\index{network|(}
```

While, at the end of this chapter, add |) like this:

```
\index{network|)}
```

This will produce an index entry such as **Network, 15-17**.

Using symbols and macros in the index

By default, `makeindex` sorts all entries alphabetically. If you want to include symbols in the index – for example, Greek letters, chemical formulas, or math symbols – you may run into sorting challenges. To handle this, the `\index` command supports a sort key. Use this key as a prefix for the entry, separated by the @ symbol, for instance:

```
\index{Gamma@$\Gamma$}
```

Using macros for index entries is generally not recommended. The macro name, including the backslash, would determine the sorting, although the macro would be expanded in the index. Imagine you've got a `\group` macro that stands for TeX Users Group, defined like this:

```
\newcommand{\group}{\TeX\ Users Group}
```

If you write the following, then the TeX Users Group entry will be treated like `\group` in the sorting and won't appear among the entries beginning with T:

```
\index{\group}
```

However, you could repair such issues by adding a sort key as a prefix, such as here:

```
\index{TeX@\group}
```

You can also indicate how words with special characters will be sorted. Here, the German word *schön*, which contains an umlaut, will be sorted like the word *schon*:

```
\index{schon@sch\"{o}n}
```

Since the symbols `|`, `@`, and `!` have special meanings within index entries, we need to quote them to print them as their original meanings. This is an example of how we can print them:

```
\index{exclamation ("!)!loud}
```

We can print the symbols `|`, `@`, and `!` in the index by quoting them, using a preceding `"`.

Referring to other index entries

Different words may stand for the same concept. In such cases, it's possible to add a cross-reference to the main phrase without a page number. Adding the `|see{entry list}` code achieves that, for example:

```
\index{wireless|see{WLAN}}
\index{WLAN}
```

As such, references don't print a page number, so their position in the text doesn't matter. You could collect them anywhere in your document.

Fine-tuning page numbers

If an index entry refers to several pages, you should emphasize a specific page number to indicate it as the primary reference. You could define a command for emphasizing as follows:

```
\newcommand{\main}[1]{\emph{#1}}
```

Then, for the index entry, add a pipe symbol followed by the command name:

```
\index{WLAN|main}
```

Now, LaTeX emphasizes the corresponding page number. Simply writing `\index{WLAN|emph}` or `\index{WLAN|texbf}` is also possible. However, defining your own macro is more consistent: remember the concept of separating form and content.

Designing the index layout

If we extend our example document with the example commands mentioned in the previous sections, `\printindex` gives us this layout, containing subentries, ranges, references, and emphasized entries:

Index

TeX Users Group sorted wrong, 17

enterprise, 9
    organization, 9
        controlling, 9
        operation, 9
        sales, 9
exclamation (!)
    loud, 17

Γ, 17

network, 15–17
    overview, 15

schön, 17

TeX Users Group sorted correctly,
        17

wireless, *see* WLAN
WLAN, *17*

Figure 8.4 – A more complex index

LaTeX comes with several built-in index styles, called `latex` (the default), `gind`, `din`, and `iso`. To use another style, specify it using the `-s` option of the `makeindex` program, for example:

```
makeindex -s iso documentname
```

If you compile after this call, the index layout changes to the following:

Index

TEX Users Group sorted wrong . 17
enterprise 9
  organization 9
    controlling 9
    operation 9
    sales 9
exclamation (!)
  loud 17
Γ 17
network 15–17
  overview 15
schön 17
TEX Users Group sorted correctly
    17
wireless . . . . *see* WLAN
WLAN *17*

Figure 8.5 – An index with the iso style

You could even define your own styles. To learn more about indexing and `makeindex`, use `texdoc` at the command prompt:

```
texdoc index
```

For more information about the `makeindex` tool, use the following command:

```
texdoc makeindex
```

Or, visit the documentation online at `https://texdoc.org/pkg/index` and `https://texdoc.org/pkg/makeindex`.

Although it might seem natural to generate the index while writing the document, it's recommended to wait until the writing is finished and then decide what should appear in the index.

Next, we'll move on to another type of list: the list of references, that is, the bibliography.

Creating a bibliography

Scientific and academic documents usually include a list of **references** or a **bibliography**. We'll work out how to typeset a bibliography and cite its entries in the text.

Using LaTeX's built-in features, we'll create a short list of references containing a book and an article by Donald E. Knuth, the creator of TeX. In our main text, we will cite both of them:

1. Create a new document as follows:

```
\documentclass{article}
\begin{document}
\section*{Recommended texts}
To study \TeX\ in depth, see \cite{DK86}.
For writing math texts, see \cite{DK89}.
\begin{thebibliography}{8}
\bibitem{DK86} D.E. Knuth, \emph{The {\TeX}book}, 1986
\bibitem{DK89} D.E. Knuth, \emph{Typesetting Concrete
Mathematics}, 1989
\end{thebibliography}
\end{document}
```

2. Click on **Typeset** and examine the output:

Recommended texts

To study TEX in depth, see [1]. For writing math texts, see [2].

References

[1] D.E. Knuth, *The TEXbook*, 1986

[2] D.E. Knuth, *Typesetting Concrete Mathematics*, 1989

Figure 8.6 – A list of references

Here, we used an environment called `thebibliography` to typeset the list of references, which is similar to a description list, as you saw in *Chapter 4, Creating Lists*. Each item on this list has a key that identifies it. For citing in the body text, we used the `\cite` command to refer to that key. Let's take a closer look at how these commands work.

Using the built-in bibliography environment

LaTeX's default environment for bibliographies has the following structure:

```
\begin{thebibliography}{widest label}
\bibitem[label]{key} author, title, year etc.
\bibitem…
…
\end{thebibliography}
```

Each item is specified using the `\bibitem` command. This command requires a mandatory argument determining the key. We may refer to this key with `\cite{key}` or `\cite{key1,key2}`. `\cite` also accepts an optional argument stating a page range, for example, `\cite[p.\,18--20]{key}`. You may choose a label by means of the optional argument of `\bibitem`. If we did not state a label, LaTeX would number the items consecutively in square brackets, as we saw in *Figure 8.6*.

Using labels, the environment could look as follows:

```
\begin{thebibliography}{Knuth89}
\bibitem[Knuth86]{DK86} D.E. Knuth, \emph{The {\TeX}book}, 1986
\bibitem[Knuth89]{DK89} D.E. Knuth, \emph{Typesetting Concrete
  Mathematics}, 1989
\end{thebibliography}
```

The corresponding output is this:

Recommended texts

To study TEX in depth, see [Knuth86]. For writing math texts, see [Knuth89].

References

[Knuth86] D.E. Knuth, *The TEXbook*, 1986

[Knuth89] D.E. Knuth, *Typesetting Concrete Mathematics*, 1989

Figure 8.7 – A list of references

As you can see, LaTeX automatically adjusted the output of `\cite` to the new labels. If you want more control over citation formatting, the `cite` package offers compressed and sorted lists of numerical citations, such as `[2,4-6]`, and further formatting options for in-text citations.

The mandatory environment argument should include the widest label for alignment. So, for instance, if you have more than 9 but fewer than 100 items, you may write two digits into the argument.

Using bibliography databases with BibTeX

Manually creating the bibliography can be tedious. Especially if you use references across several documents, it would be better to use a database and let a program generate the bibliography for you. This sounds more complicated than it actually is. Let's try it.

We'll create a separate database file containing the references from our previous example. We'll modify our example to use that database. To make this work, we have to call the external program **BibTeX**:

1. Create a new document. Begin by writing the entry for **TeXbook**:

```
@book{DK86,
author = "D.E. Knuth",
title = "The {\TeX}book",
publisher = "Addison Wesley",
year = 1986
}
```

2. Next, add the article, where we'll specify even more fields:

```
@article{DK89,
author = "D.E. Knuth",
title = "Typesetting Concrete Mathematics",
journal = "TUGboat",
volume = 10,
number = 1,
pages = "31--36",
month = apr,
year = 1989
}
```

3. Save the file and name it `example.bib`. Open our example document and modify it as follows:

```
\documentclass{article}
\begin{document}
\section*{Recommended texts}
To study \TeX\ in depth, see \cite{DK86}. For writing math texts,
see \cite{DK89}.
\bibliographystyle{alpha}
\bibliography{example}
\end{document}
```

4. Click **Typeset** one time with pdfLaTeX. If you're using TeXworks, choose **BibTeX** instead of **pdfLaTeX**, present in the drop-down box next to the **Typeset** button, and then click on **Typeset**. If you are writing with another editor, use its BibTeX option or type at the command prompt in the document directory as follows, without a file name extension:

```
bibtex documentname
```

5. Click on **Typeset** twice with pdfLaTeX. Here's the result:

Recommended texts

To study TEX in depth, see [Knu86]. For writing math texts, see [Knu89].

References

[Knu86] D.E. Knuth. *The TEXbook*. Addison Wesley, 1986.

[Knu89] D.E. Knuth. Typesetting concrete mathematics. *TUGboat*, 10(1):31–36, April 1989.

Figure 8.8 – A bibliography based on a database file

In this setup, we created a text file containing all the bibliography entries. Our document uses the `alpha` style, which sorts entries by the authors' names and uses a shortcut consisting of `the author` and `year` keys as the label. We told LaTeX to load the bibliography file called `example`. The `.bib` extension is assumed automatically.

Then, we called the external program **BibTeX**. This program knows from the example `.tex` file that `example.bib` needs to be translated. Thus, out of this `.bib` file, it creates a `.bbl` file containing a LaTeX `thebibliography` environment and the final entries.

Finally, we had to compile twice to ensure that all the cross-references are correct.

Although this process involves a few extra steps to generate the bibliography, there are also advantages: we don't need to fine-tune each entry, we can easily switch between styles, and we can then reuse the `.bib` file.

Now, let's look at the `.bib` file format more closely.

Looking at the BibTeX entry fields

BibTeX supports various entry types, as we saw with the `book` and `article` entries. Furthermore, these entries contain fields such as `author`, `title`, and `year`. Let's first look at the supported fields, and afterward, we'll discuss the different kinds of entries.

Here's a list of the standard fields. Some fields are common, some are rarely used; we'll list them in alphabetical order, for completeness:

- `address`: The address of the publisher
- `annote`: An annotation not used by the standard bibliography styles
- `author`: The name(s) of the author(s)
- `booktitle`: The title of a book if you cite a part of it; you can also use the `title` field instead
- `chapter`: A chapter number
- `crossref`: The key of the database entry being cross-referenced
- `edition`: The edition (first, second, and so on) of a book; it's commonly in uppercase
- `editor`: The name(s) of the editor(s)
- `howpublished`: The way of publishing; the first word should be in uppercase
- `institution`: This could be a sponsoring institution
- `journal`: A journal name; you may use standard abbreviations
- `key`: Used for alphabetizing, cross-referencing, and labeling if the author's information is missing; don't confuse it with the key used in the `\cite` command
- `month`: The month in which the work was published; a three-letter abbreviation is often used
- `note`: Any additional helpful information; again, the first word should be in uppercase
- `number`: The number of a journal or another kind of work in a series
- `organization`: This can be a sponsoring organization

- `pages`: A page number or range of page numbers, such as 12-18 or 22+
- `publisher`: The name of the publisher
- `school`: This could be the name of the school where the document was written
- `series`: The name of a series of books or its number in a multi-volume set
- `title`: The title of the work
- `type`: The type of the publication
- `volume`: The volume of a journal or multi-volume book
- `year`: The year of the publication

You can read the BibTeX documentation by typing `texdoc bibtex` at the command line or visiting `https://texdoc.org/pkg/bibtex`.

Citing internet resources

Nowadays, it's common to cite web pages and other internet sources. To put internet addresses into BibTeX fields, use the `\url` command of the `url` or `hyperref` package, for example:

```
howpublished = {\url{https://latex.org}}
```

Some styles' bibliographies include a `URL` field that can be used to refer to internet addresses.

Understanding BibTeX entry types

Usually, you start by choosing the appropriate entry type you want to add, and then you fill in the fields. Different types may support various fields. Some fields are required, some are optional and may be omitted, and some are ignored when the style doesn't support them.

Usually, the name of the entry tells you its meaning. These are the standard BibTeX entry types and their required and optional fields:

Type	Required fields	Optional fields
article	author, title, journal, year	volume, number, pages, month, note
book	author or editor, title, publisher, year	volume or number, series, address, edition, month, note
booklet	title	author, howpublished, address, month, year, note
conference	author, title, booktitle, year	editor, volume or number, series, pages, address, month, organization, publisher, note
manual	title	author, organization, address, edition, month, year, note
mastersthesis	author, title, school, year	type, address, month, note
misc	none	author, title, howpublished, month, year, note
phdthesis	author, title, school, year	type, address, month, note
proceedings	title, year	editor, volume or number, series, address, month, organization, publisher, note
techreport	author, title, institution, year	type, number, address, month, note
unpublished	author, title, note	month, year

Figure 8.9 – BibTeX entry types and fields

Have a look at the BibTeX reference for more details by typing the following at the command prompt:

```
texdoc bibtex
```

Alternatively, you can visit `https://texdoc.org/pkg/bibtex`.

If no other entry fits, choose the `misc` type. It doesn't matter whether you use capitals or small letters in a type; `@ARTICLE` is understood the same as `@article`. As the example shows, entries have the following form:

```
@entrytype{keyword,
fieldname = {field text},
otherfieldname = {other field text},
...
}
```

Enclose the `field text` in braces. You can also use straight quotes instead, as in `"field text"`. For numbers, you may even omit the braces. While entries are separated by a comma, never put a comma after the last field. Otherwise, BibTeX assumes another field must follow and may report an error.

Some styles change capitalization, which might result in undesired lowercase letters. To protect letters or words from becoming lowercase, put additional braces around them. Preferably, do it around a word rather than just the letter to keep ligatures and improve spacing. For example, `{WAL}` looks better than `{W}AL`, because in a standard text flow, LaTeX moves an **A** closer to a preceding **W**. Separating with braces interferes with LaTeX's micro-typographic improvements.

Choosing the bibliography style

BibTeX offers several standard styles:

- `plain`: Uses Arabic numbers for the labels, sorted alphabetically by the authors' names. The number is written in square brackets, which also appear with `\cite`.
- `unsrt`: There's no sorting. All entries appear in the order they were cited in the text. Apart from this, it looks like the `plain` style.
- `alpha`: Sorting is based on authors' names; the labels are shortcuts consisting of the author's name and the year of publication. Square brackets are used here as well.
- `abbrv`: This works like `plain`, but abbreviates first names and other field entries.

Choose the style after \begin{document} but before \bibliography. You may write \bibliographystyle right before \bibliography to keep it together.

There are more styles available in TeX distributions and on the internet. For instance, the natbib package provides styles and the capability to cite in a nice **author-year** scheme. This package further adds some fields, such as ISBN, ISSN, and URL.

You could give the natbib package a try and use its plainnat, abbrvnat, and unsrtnat styles, for instance:

```
\usepackage{natbib}
\bibliographystyle{plainnat}
```

The natbib package reimplemented the \cite command and offers variations to it, with the main purpose of supporting author-year citations. It works with most other available styles. It introduces the citation command, \citet, for textual citations, and the \citep command for parenthetical citations. There are starred variants that print the full author list, and optional arguments that allow adding text before and after the list.

Check out the documentation if you would like to take advantage of this excellent package. As usual, type texdoc natbib at the command line or visit https://texdoc.org/pkg/natbib.

The biblatex package is a complete reimplementation of BibTeX's bibliographic functionality. While BibTeX relies on a separate, stack-based style language, biblatex is native to LaTeX, which makes bibliography styles easier to customize and extend. You can find the documentation at https://texdoc.org/pkg/biblatex.

It works closely with the modern **biber** program, which replaces the traditional BibTeX program; see https://texdoc.org/pkg/biber for documentation. In the *LaTeX Cookbook, 8, Producing Contents, Indexes, and Bibliographies,* you can find a well-explained, step-by-step example using biblatex and biber.

Listing references without citing

By default, BibTeX includes only references from the database that are cited in the text and prints them. However, you may specify keys for references, which should nevertheless appear. Just write the following for a single reference:

```
\nocite{key}
```

Or, write the following to list the complete database:

```
\nocite{*}
```

Make sure to remove `\nocite{*}` in the final version of the document if you don't want to have references in the bibliography that you never cited in the document.

Now that we know how to create such tables of contents, lists of objects, indexes, and bibliographies, let's have a final look at how to customize them.

Changing the headings

As shown in *Figure 8.2*, you can easily change the **Contents** heading if you don't like it. LaTeX stores the text of the heading in the `\contentsname` text macro. So, you can redefine it as follows:

```
\renewcommand{\contentsname}{Table of Contents}
```

Here's a list of such macros and their default values:

- `\contentsname`: Contents
- `\listfigurename`: List of figures
- `\listtablename`: List of tables
- `\bibname`: Bibliography (in the `book` and `report` classes)
- `\refname`: References (in the `article` class)
- `\indexname`: Index

Furthermore, here's a list of other macros for names used by LaTeX, with their default values:

- `\figurename`: Figure
- `\tablename`: Table
- `\partname`: Part
- `\chaptername`: Chapter
- `\abstractname`: Abstract
- `\appendixname`: Appendix

This is not really surprising! Using name macros is especially useful when you write in another language. For instance, the `babel` package takes a language option and redefines all those name macros according to the chosen language.

However, they are also helpful when choosing abbreviations, such as **Fig.,** or different phrasing, such as **Appendices** instead of **Appendix**.

Adding a glossary

A **glossary** is a structured list of terms with their definitions. It's commonly used to explain technical vocabulary, abbreviations, and symbols in theses, scientific papers, technical manuals, and books, especially when terms recur throughout the text. Closely related are lists of **acronyms** and **symbols**, which serve the same purpose: helping readers find explanations quickly without interrupting the flow of the document.

In LaTeX, the recommended modern solution is the `glossaries` package. It's powerful and flexible, handling everything from a short list of abbreviations to large, automatically generated glossaries in complex documents. Some highlights of this package include support for glossaries, acronyms, and symbols in a single framework; automatic sorting and formatting; fine control over entry display; integration with indexing tools; and support for multilingual documents.

The package is exceptionally well documented. It comes with a beginner's guide, a detailed user manual, and an author FAQ, all available at `https://ctan.org/pkg/glossaries`.

Because the accompanying guides are so thorough, we won't go into the details here. The point is: if you need a glossary or a list of symbols, this package is highly recommended.

Summary

In this chapter, we covered many types of lists. Specifically, we learned how to generate and customize the table of contents and lists of figures and tables, produce an index pointing to relevant information for keywords and phrases, and create bibliographies, both manually and using a bibliography database.

These lists serve as navigation aids, guiding readers to the information they seek. They aren't merely for listing and summarizing. Sometimes, there's a strange requirement to list the table of contents within itself. If you are not sure about a design or a requirement, have a look at a good book in your particular field to see what exemplary tables of contents, lists, and indexes might look like.

You can find related examples in the *LaTeX Cookbook, 8, Producing Contents, Indexes, and Bibliographies*, with code examples available on the book's website: `https://latex-cookbook.net/chapter-08`.

In the next chapter, we'll take a closer look at scientific writing.

Writing Math Formulas

At the beginning of this book, in *Chapter 1, Getting Started with LaTeX*, we promised that LaTeX offers excellent quality for mathematical typesetting. Now, it's time to show this. By the end of this chapter, you'll be able to write clear and elegant mathematical text.

To make full use of LaTeX's math capabilities, we'll work through the following topics:

- Writing basic formulas
- Typesetting multi-line formulas
- Exploring the wealth of math symbols
- Building math structures

It's quite an undertaking, so let's get started.

Writing basic formulas

LaTeX provides three writing modes:

- **Paragraph mode**: Text is typeset as a sequence of words arranged into lines, paragraphs, and pages. That's what we've used throughout the previous chapters.
- **Left-to-right mode**: Text is set as a continuous sequence of words without line breaks. For example, the argument of the `\mbox` command is typeset in this mode, so we can use it to prevent a phrase from being hyphenated.

- **Math mode**: Here, LaTeX treats letters as mathematical symbols. They will appear in italics, as is common practice for variables. Many symbols are available only in math mode, such as roots, summation signs, relation symbols, math accents, arrows, and various delimiters, such as parentheses, brackets, and braces. LaTeX ignores normal space characters between letters and symbols in math mode. Instead, the spacing is determined by the type of symbol, so spacing around relation signs is different from the spacing around opening or closing delimiters. Any mathematical expression must be written in this mode.

Now, we'll enter the math mode for the first time.

Our first math text shall deal with the solutions of quadratic equations. We will typeset formulas with constants and variables, use superscripts for the squares and subscripts for the solutions, and include a root symbol. Finally, we will add cross-references to formulas. That's quite a bit to cover, so let's break it down into the following steps:

1. Start a new document. For now, we don't need any packages:

   ```
   \documentclass{article}
   \begin{document}
   \section*{Quadratic equations}
   ```

2. State the quadratic equation with its conditions. Use an `equation` environment for the formula. Surround small pieces of math within text using `\(` and `\)`:

   ```
   The quadratic equation
   \begin{equation}
     \label{quad}
     ax^2 + bx + c = 0,
   \end{equation}
   where \( a, b \) and \( c \) are constants and
   \( a \neq 0 \), has two solutions for the variable
   \( x \):
   ```

3. Write another equation for the solutions. The command for a square root is `\sqrt`, and for a fraction, use the `\frac` command:

   ```
   \begin{equation}
     \label{root}
     x_{1,2} = \frac{-b \pm \sqrt{b^2-4ac}}{2a}.
   \end{equation}
   ```

4. Let's introduce the discriminant and discuss the case when it's zero. To get an unnumbered displayed equation, we surround the formula with `\[` and `\]`:

```
If the \emph{discriminant} \( \Delta \) with
\[
  \Delta = b^2 - 4ac
\]
is zero, then the equation (\ref{quad}) has a double solution:
(\ref{root}) becomes
\[
  x = - \frac{b}{2a}.
\]
\end{document}
```

5. Compile the document. The equation references are unresolved in the first run and will appear as `(?)`. Compile again to let LaTeX resolve them, then look at the final output:

Quadratic equations

The quadratic equation

$$ax^2 + bx + c = 0, \tag{1}$$

where a, b and c are constants and $a \neq 0$, has two solutions for the variable x:

$$x_{1,2} = \frac{-b \pm \sqrt{b^2 - 4ac}}{2a}. \tag{2}$$

If the *discriminant* Δ with

$$\Delta = b^2 - 4ac$$

is zero, then the equation (1) has a double solution: (2) becomes

$$x = -\frac{b}{2a}.$$

Figure 9.1 – An example of a mathematical text

As mentioned in *Chapter 1, Getting Started with LaTeX*, writing formulas often feels similar to programming. We build expressions from commands; some take arguments, such as those for roots and fractions, while others are simple symbol commands, such as those for Greek letters. Most of these symbols work only in math environments, not in regular text. This chapter will help you master it, and the results will be well worth the effort.

The `equation` environment created a displayed formula. LaTeX centered it horizontally and added some vertical space above and below. Furthermore, it numbered these formulas consecutively.

However, `\[ ... \]` and `\( ... \)` are also, in reality, environments. Let's look at them more closely in the next sections.

Embedding math expressions within text

LaTeX provides the `math` environment for in-text formulas:

```
\begin{math}
  expression
\end{math}
```

Since it can be tedious to write this environment for each small expression or symbol, LaTeX offers a shorter form that behaves the same:

```
\(
  expression
\)
```

You can place it all on one line as well, such as `\( expression \)`.

A third way is to use a shortcut from TeX: `$expression$`. A disadvantage of the latter is that the opening and closing symbols are the same, which may easily lead to mistakes. Still, it's much quicker to type, which is why it remains popular among LaTeX users, and it's perfectly okay to use it.

Inline formulas save space and help keep explanations flowing, so they are recommended for short mathematical expressions within the text.

Displaying formulas

For centered, displayed formulas, LaTeX provides the `displaymath` environment:

```
\begin{displaymath}
  expression
\end{displaymath}
```

With this environment, when the paragraph ends, some vertical space follows, then LaTeX displays the centered formula, and adds vertical space afterward. Because this environment already handles spacing, avoid leaving empty lines before or after it, as this would create additional vertical space from the extra paragraph breaks.

There's also a shorter form for this environment. Here, it's with square brackets, instead of parentheses:

```
\[
  expression
\]
```

Placing the `\[` and `\]` shortcuts on their own lines makes the code easier to read.

In older tutorials and on the internet, you may encounter the `TeX form $$...$$` for displayed formulas. Do not use this with LaTeX, as it leads to incorrect vertical spacing.

Such displayed formulas stand out more because they are centered and surrounded by additional space. Choose the style that best supports the readability of your text.

For the rest of this chapter, all pieces of code will use math mode. We will either explicitly enter a `math` environment or assume we are already in math mode for short code snippets.

Numbering equations

Equations and formulas, in general, may be numbered. However, this applies only to displayed formulas. The `equation` environment handles this:

```
\begin{equation}
  \label{key}
  expression
\end{equation}
```

It looks similar to `displaymath`, but it's numbered this time. The number will be displayed in parentheses on the right-hand side of the equation, as shown in *Figure 9.1*. You can use the key to cross-reference the equation with the `\ref` command, as we did in the previous chapter. If you don't need a reference, you can omit the `\label` command. In general, it makes sense to number only the equations that you plan to reference.

If you load the `amsmath` package, which we'll use later, you can write unnumbered equations with the starred `equation*` environment.

Adding subscripts and superscripts

Since **exponents** and **indices** are used so often, LaTeX provides concise commands for them. An underscore, _, creates a **subscript**, often used as an index:

```
{expression}_{subscript}
```

A caret, ^, produces a **superscript** used for exponents:

```
{expression}^{superscript}
```

As we see here, we use braces to define the relevant part of the expression. In the case of single letters, numerals, or symbols, you can omit the braces.

Subscripts and superscripts may be nested. If you use both subscripts and superscripts in the same expression, the order of ^ and _ doesn't matter. Let's look at an example:

```
\[ x_1^2 + x_2^2 = 1, \quad 2^{2^x} = 64 \]
```

This gives us the following output:

$$x_1^2 + x_2^2 = 1, \quad 2^{2^x} = 64$$

Figure 9.2 – Subscripts and superscripts

Notice that the higher-level exponent is smaller than the lower one. When we nest subscripts or superscripts, the inner font size becomes smaller.

The name of the `\quad` command comes from traditional metal typesetting, where a square piece was inserted to add horizontal space. `\quad` creates a small space roughly the width of the letter *M*. The `\qquad` command produces twice that amount. Both commands are useful for quick spacing adjustments.

Using operators

Trigonometric, logarithmic, and other analytic and algebraic **functions** are commonly written with upright Roman letters, in contrast to variables, which are written in italics. Simply typing `log` would otherwise look like a product of the three variables: `l`, `o`, and `g`. LaTeX provides commands for many common functions, often called **operators**. Here's an alphabetical list of the predefined ones: `\arccos`, `\arcsin`, `\arctan`, `\arg`, `\cos`, `\cosh`, `\cot`, `\coth`, `\scs`, `\deg`, `\det`, `\dim`, `\exp`, `\gcd`, `\hom`, `\inf`, `\ker`, `\lg`, `\lim`, `\liminf`, `\limsup`, `\ln`, `\log`, `\max`, `\min`, `\Pr`, `\sec`, `\sin`, `\sinh`, `\sup`, `\tan`, and `\tanh`.

You can write the modulo function in two ways: either use \bmod for a binary relation or use \pmod{argument} for a modulo expression in parentheses.

Some operators take subscripts, which appear below the operator in displayed formulas:

```
\[ \lim_{n=1, 2, \ldots} a_n \qquad \max_{x<X} x \]
```

The output is as follows:

$$\lim_{n=1,2,\ldots} a_n \qquad \max_{x<X} x$$

Figure 9.3 – Operators with subscripts

Superscripts would be set above the operator.

When operators are used inline within text, it's slightly different:

```
Within text, we have \( \lim_{n=1, 2, \ldots} a_n \)
and \( \max_{x<X} x \).
```

The output is now as follows:

Within text, we have $\lim_{n=1,2,\ldots} a_n$ and $\max_{x<X} x$.

Figure 9.4 – Subscripts

That's to avoid excessive line spacing.

In addition, *Figures 9.30* and *9.31* show the positioning and size of subscripts and superscripts for operators.

In *Chapter 10, Writing Advanced Mathematics*, in *LaTeX Cookbook*, you can read how to define your own operators and how to fine-tune subscripts and superscripts for perfect alignment.

Typesetting roots

Our first example in this chapter contained a **square root** written as \sqrt{value}. Because this command also supports **higher-order roots**, it accepts an optional argument specifying the order. The complete definition is as follows:

```
\sqrt[order]{value}
```

Roots may be nested. We can see it in this example:

```
\sqrt[64]{x} = \sqrt{\sqrt{\sqrt{\sqrt{\sqrt{\sqrt{x}}}}}}
```

This produces the following:

$$\sqrt[64]{x} = \sqrt{\sqrt{\sqrt{\sqrt{\sqrt{\sqrt{x}}}}}}$$

Figure 9.5 – Nested roots

LaTeX automatically adjusts the size of the root symbol to the height and the width of the `value` expression. This is why outer roots appear larger than inner roots.

Writing fractions

For inline formulas, you may just use a slash, `/`, to write **fractions**, such as `\( (a+b)/2 \)`. For larger fractions, use the `\frac` command:

```
\frac{numerator}{denominator}
```

Here is an example:

```
\[ \frac{n(n+1)}{2} \quad \frac{\frac{\sqrt{x}+1}{2}-x}{y^2} \]
```

The output is as follows:

$$\frac{n(n+1)}{2} \quad \frac{\frac{\sqrt{x}+1}{2}-x}{y^2}$$

Figure 9.6 – Fractions and nested fractions

LaTeX automatically adjusts the separation line to match the width of the numerator and denominator.

Writing Greek letters

Mathematicians often use Greek letters, for example, to denote **constants**. To get a lowercase Greek letter, write the name as a command with a backslash. Here are the lowercase Greek letters with their corresponding LaTeX commands:

α \alpha	ζ \zeta	λ \lambda	π \pi	ϕ \phi
β \beta	η \eta	μ \mu	ρ \rho	χ \chi
γ \gamma	θ \theta	ν \nu	σ \sigma	ψ \psi
δ \delta	ι \iota	ξ \xi	τ \tau	ω \omega
ϵ \epsilon	κ \kappa	o o	υ \upsilon	

Figure 9.7 – Lowercase Greek letters

For some letters, variants are available:

ε `\varepsilon`	ϖ `\varpi`	ς `\varsigma`
ϑ `\vartheta`	ϱ `\varrho`	φ `\varphi`

Figure 9.8 – Alternative variants for some Greek letters

As the omicron looks like the letter o, there's no command for it. The same applies to most uppercase Greek letters, which look identical to their Roman counterparts. For example, there's no `\Alpha` or `\Beta` commands; just type `A` or `B` instead. The uppercase Greek letters that differ from Roman letters can be written as follows:

Γ `\Gamma`	Λ `\Lambda`	Σ `\Sigma`	Ψ `\Psi`
Δ `\Delta`	Ξ `\Xi`	Υ `\Upsilon`	Ω `\Omega`
Θ `\Theta`	Π `\Pi`	Φ `\Phi`	

Figure 9.9 – Uppercase Greek letters

You can see that lowercase Greek letters are typeset in italics, and uppercase Greek letters are written upright. This convention reflects traditional mathematical typesetting. The frugality of using only italic lowercase letters and a limited number of upright Greek letters stems from space limitations in character tables in the early days of TeX.

If you want to have upright Greek letters, you may add `\usepackage{upgreek}` and then use the following commands:

α `\upalpha`	ζ `\upzeta`	λ `\uplambda`	π `\uppi`	ϕ `\upphi`
β `\upbeta`	η `\upeta`	μ `\upmu`	ρ `\uprho`	χ `\upchi`
γ `\upgamma`	θ `\uptheta`	ν `\upnu`	σ `\upsigma`	ψ `\uppsi`
δ `\updelta`	ι `\upiota`	ξ `\upxi`	τ `\uptau`	ω `\upomega`
ϵ `\upepsilon`	κ `\upkappa`	ο `\mathrm{o}`	υ `\upupsilon`	

Figure 9.10 – Upright lowercase Greek letters

The following additional variants are also available:

ε `\upvarepsilon`	ϖ `\upvarpi`	ς `\upvarsigma`
ϑ `\upvartheta`	ϱ `\upvarrho`	φ `\upvarphi`

Figure 9.11 – Alternative upright variants for some Greek letters

The upright Greek letters are taken from the Euler font and not from the default Computer Modern fonts.

Writing script letters

For the 26 uppercase letters A, B, C, ..., Z, there's a **calligraphic** shape, produced by `\mathcal`:

```
\[ \mathcal{A}, \mathcal{B}, \mathcal{C}, \ldots, \mathcal{Z} \]
```

This is how they look:

$$\mathcal{A}, \mathcal{B}, \mathcal{C}, \ldots, \mathcal{Z}$$

Figure 9.12 – Calligraphic letters

Several packages offer alternative calligraphic fonts, such as `zapfino` and `xits`.

Producing an ellipsis

You already know that the `\ldots` command produces a low **ellipsis**. It also works in math mode. We use the low ellipsis mainly between letters and commas. Between operation and relation symbols, a centered ellipsis is more common. Also, a matrix may require a diagonal or a vertical ellipsis. Here's how we can produce them:

$\ldots$	`\ldots`	$\ddots$	`\ddots`
$\cdots$	`\cdots`	$\vdots$	`\vdots`

Figure 9.13 – Ellipsis in various positions

For a diagonal ellipsis in the opposite direction, you can write `\reflectbox{$\ddots$}`. The `\reflectbox` command requires `\usepackage{graphicx}` in your preamble.

Changing the font, style, and size

In *Chapter 2, Formatting Text and Creating Macros*, we looked at ways to modify the text font. In math mode, we can use additional commands to change the font style, as follows:

Command	Used package	Example
`\mathrm{...}`		$\mathrm{roman\ 123}$
`\mathit{...}`		$\mathit{italic\ 123}$
`\mathsf{...}`		$\mathsf{sans-serif\ 123}$
`\mathbb{...}`	amsfonts	$\mathbb{ABC}$
`\mathbbm{...}`	bbm	$\mathbbm{CRQZ1}$
`\mathds{...}`	dsfont	$\mathds{CRQZ1}$
`\mathfrak{...}`	eufrak	$\mathfrak{ABC\ 123}$
`\mathnormal{...}`		$\mathnormal{normal}$

Figure 9.14 – Math font commands

For example, once you add `\usepackage{dsfont}` to your document preamble, you can write `\mathds{Z}` to get a double-stroke letter *Z*.

Although letters in math mode are italicized by default, LaTeX treats them as separate symbols, resulting in a different spacing than in an italic word. For instance, in math mode, `fi` may represent the product of the `f` and `i` variables but not the `fi` ligature. Compare these two versions:

```
\(Definition\) and \textit{Definition}
```

This gives us the following:

$$Definition \text{ and } \textit{Definition}$$

Figure 9.15 – Plain math writing versus italic text

The version on the right is clearly better.

Also, `\textit` typesets its arguments using the italic math font, not the standard text font. For text within formulas, please read the *Inserting text into formulas* section later in this chapter.

To make an entire math expression bold, use the `\boldmath` declaration before the expression, that is, outside math mode. The `\unboldmath` declaration switches back to the standard typeface. The latter is also used outside math mode.

To make only a part of a formula bold, you can switch to left-to-right mode with the `\mbox` command and use `\boldmath` in its argument.

Four different **math styles** are available, which determine the typesetting and font size. You can use these commands within math mode:

- `\textstyle` enforces inline style for symbol sizes and positioning of subscripts and superscripts, even when used in a displayed formula
- `\displaystyle` treats symbol sizes, subscripts, and superscripts as in a displayed formula, even when used inline in text
- `\scriptstyle` switches to a smaller font size, such as for subscripts and superscripts
- `\scriptscriptstyle` uses a much smaller font size for the nested script style

`\textstyle` differs from `\displaystyle` in mainly two ways: with `\textstyle`, variable-sized symbols are smaller, and subscripts and superscripts are usually placed beside the expression instead of below and above, respectively. Otherwise, the font size is the same.

LaTeX switches the style automatically. If you write a simple exponent, it will be typeset in script style with a smaller font size.

You may force the desired style using one of the four commands listed here. So, for instance, you can insert `\displaystyle` into a formula, so even within the text, it would appear like in a displayed formula: bigger fraction and bigger sum signs. Furthermore, subscripts are placed below, and superscripts are placed above. Note, this increases the line spacing.

Customizing displayed formulas

Two document class options can change how displayed formulas appear:

- `fleqn` for **flush left equations**: LaTeX will align all displayed formulas at the left margin
- `leqno` for **left equation numbers**: all numbered formulas will get the numbers on the left side instead of the right

Equations and formulas are often part of a larger structure rather than standing alone. We may run into situations such as the following:

- A formula is too long to fit on one line
- Several formulas need to be listed one after another
- An equation must be transformed step by step
- A chain of inequalities spans multiple lines
- Several formulas must be aligned at their relation symbols

We may also encounter situations where we need to write multi-line equations, often with some form of alignment. The `amsmath` package provides specialized environments for almost all such situations, which will be the topic of our next section.

Typesetting multi-line formulas

We'll use the `amsmath` package to typeset a very long formula and a system of equations:

1. Start a new document. Here, we use A6 paper size to work with a smaller text width:

```
\documentclass{article}
\usepackage[a6paper]{geometry}
```

2. Load the `amsmath` package, and begin the document:

```
\usepackage{amsmath}
\begin{document}
```

3. Use the `multline` environment to spread a long equation over three lines. End each line with a double backslash, \\, except the last one:

```
\begin{multline}
  \sum = a + b + c + d + e \\
           + f + g + h + i + j \\
           + k + l + m + n
\end{multline}
\end{document}
```

4. Compile and look at the equation:

$$\begin{multline} \sum = a + b + c + d + e \\ + f + g + h + i + j \\ + k + l + m + n \qquad (1) \end{multline}$$

Figure 9.16 – A formula spanning three lines

5. Now, we handle a system of equations. Use the `gather` environment to add these equations. Again, end all lines with \\, except the last one:

```
\begin{gather}
  x + y + z = 0 \\
      y - z = 1
\end{gather}
```

6. Compile again, and look at the equations, which appear horizontally centered to each other:

$$x + y + z = 0 \qquad (2)$$
$$y - z = 1 \qquad (3)$$

Figure 9.17 – A system with two equations

7. Equation systems are often aligned at the equal sign. Let's do this. Use the ampersand symbol, &, to mark the point where we wish to align:

```
\begin{align}
  x + y + z &= 0 \\
      y - z &= 1
\end{align}
```

8. Compile again; now the equations are aligned as desired:

$$
\begin{aligned}
x + y + z &= 0 \qquad (4)\\
y - z &= 1 \qquad (5)
\end{aligned}
$$

Figure 9.18 – A system with two aligned equations

Because we loaded the `amsmath` package, we now have access to several multi-line math environments. Each line in such an environment is ended by `\\`, except the last one. Otherwise, if we add `\\` to the last line, LaTeX would assume a new line and number it, even if the line is empty.

The alignment depends on the environment. Here's a list of the `amsmath` multi-line environments:

- `multline`: The first line is left-aligned, the last line is right-aligned, and all others are centered
- `gather`: Each line is centered
- `align`: Use `&` to mark a symbol where you want to align the formulas; use another `&` symbol to end a column if you need several aligned columns
- `flalign`: This is similar to `align`, but the columns are flushed to the left and the right margin, respectively
- `alignat`: This allows alignment at several places, each of which must be marked with `&`
- `split`: This is similar to `align`, but can be used within another math environment
- `aligned`, `gathered`, and `alignedat`: These are used for an aligned block within a `math` environment, which can either be displayed math or inline math

The last environments, including `split`, are often placed inside an outer `equation` environment so that several lines share a single equation number.

Numbering can be adjusted as follows.

Numbering lines in multi-line formulas

In **multi-line** math environments, each line is numbered just like a regular equation. If you want to suppress the number of a specific line, place the `\notag` command before the end of the line. It's often best to number only the lines you plan to reference.

If you prefer a particular style of numbering, such as a symbol or name as a **tag** for a formula, you can use the `\tag` command, such as `\tag{$\star$}` to mark it with a star, or `\tag{name}` for tagging it as `(name)`.

If you would like to avoid numbering altogether, use a starred version, such as `align*` or `gather*`.

Inserting text into formulas

To place regular text into a formula, standard LaTeX provides the `\mbox{text}` command. The `amsmath` package adds more flexible options:

- `\text{words}` inserts words in regular text font within a math formula and adjusts its size to the current math style. `\text` produces smaller text within subscripts or superscripts.
- `\intertext{text}` suspends the formula, prints the text as its own paragraph, then resumes the multi-line formula while preserving the alignment. This is useful for longer annotations.

These commands are helpful when you want to include text within `math` environments.

Let's now look into mathematical symbols.

Exploring the wealth of math symbols

Let's go beyond writing variables and basic math operators. Mathematical writing often requires a wide range of symbols: relation signs, unary and binary operators, function-like operators, sum and integral symbols, arrows, and many others. LaTeX and additional packages provide thousands of symbols for these purposes.

In this section, we'll look at a selection of math symbols and the commands that produce them. We'll cover many of the symbols available in standard LaTeX; the `latexsym` package adds a few more. Further packages, such as the `amssymb` package, offer an even larger collection.

Binary operation symbols

Along with plus and minus, LaTeX supports a variety of other **binary operators**:

Standard LaTeX			
$\amalg$ \amalg	$\circ$ \circ	$\ominus$ \ominus	$\star$ \star
$\ast$ \ast	$\cup$ \cup	$\oplus$ \oplus	$\times$ \times
$\bigcirc$ \bigcirc	$\dagger$ \dagger	$\oslash$ \oslash	$\triangleleft$ \triangleleft
$\bigtriangledown$ \bigtriangledown	$\ddagger$ \ddagger	$\otimes$ \otimes	$\triangleright$ \triangleright
$\bigtriangleup$ \bigtriangleup	$\diamond$ \diamond	$\pm$ \pm	$\uplus$ \uplus
$\bullet$ \bullet	$\div$ \div	$\setminus$ \setminus	$\vee$ \vee
$\cap$ \cap	$\mp$ \mp	$\sqcap$ \sqcap	$\wedge$ \wedge
$\cdot$ \cdot	$\odot$ \odot	$\sqcup$ \sqcup	$\wr$ \wr
latexsym			
$\unlhd$ \unlhd	$\unrhd$ \unrhd	$\rhd$ \rhd	$\lhd$ \lhd

Figure 9.19 – Binary operation symbols

You need to include \usepackage{latexsym} in your document preamble to use the symbols in the last row.

Relation symbols

Values of expressions might be equal, in which case you just need an equal sign, but many other **relations** are possible. For example, objects may be congruent, parallel, or they might have any other relation:

Standard LaTeX			
$\approx$ \approx	$\equiv$ \equiv	$\prec$ \prec	$\succ$ \succ
$\asymp$ \asymp	$\frown$ \frown	$\preceq$ \preceq	$\succeq$ \succeq
$\bowtie$ \bowtie	$\mid$ \mid	$\propto$ \propto	$\vdash$ \vdash
$\cong$ \cong	$\models$ \models	$\sim$ \sim	
$\dashv$ \dashv	$\parallel$ \parallel	$\simeq$ \simeq	
$\doteq$ \doteq	$\perp$ \perp	$\smile$ \smile	
latexsym			
$\Join$ \Join			

Figure 9.20 – Binary relation symbols

You can negate any relation by inserting \not before it. So, for *not equivalent*, use \not \equiv to create a crossed-out \equiv symbol.

Inequality relation symbols

If expressions are not equal, we can express that in several ways. The simplest are the basic < and > symbols, but there are also relations such as less-than-or-equal and greater-than-or-equal:

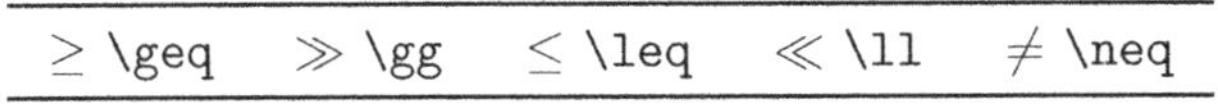

$\geq$ \geq	$\gg$ \gg	$\leq$ \leq	$\ll$ \ll	$\neq$ \neq

Figure 9.21 – Inequality relation symbols

Here, \neq looks precisely like what you get when you write \not=, as in the previous section.

Subset and superset symbols

LaTeX provides many symbols for describing how **sets** relate to one another:

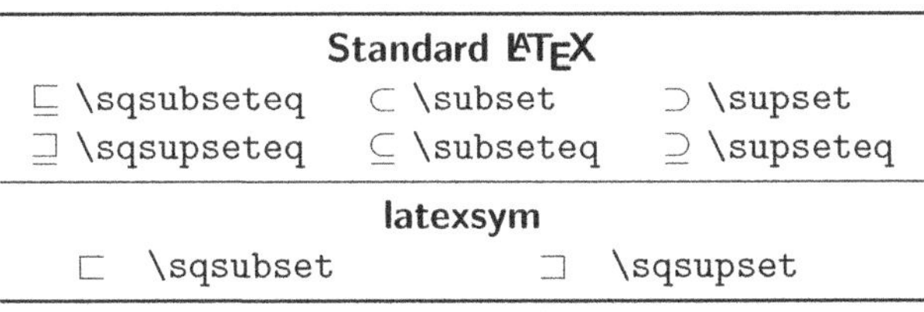

Standard LaTeX		
$\sqsubseteq$ \sqsubseteq	$\subset$ \subset	$\supset$ \supset
$\sqsupseteq$ \sqsupseteq	$\subseteq$ \subseteq	$\supseteq$ \supseteq
latexsym		
$\sqsubset$ \sqsubset	$\sqsupset$ \sqsupset	

Figure 9.22 – Subset and superset symbols

Again, you can use \not to negate such a set relation.

Arrows

LaTeX provides a wide range of **arrow symbols**:

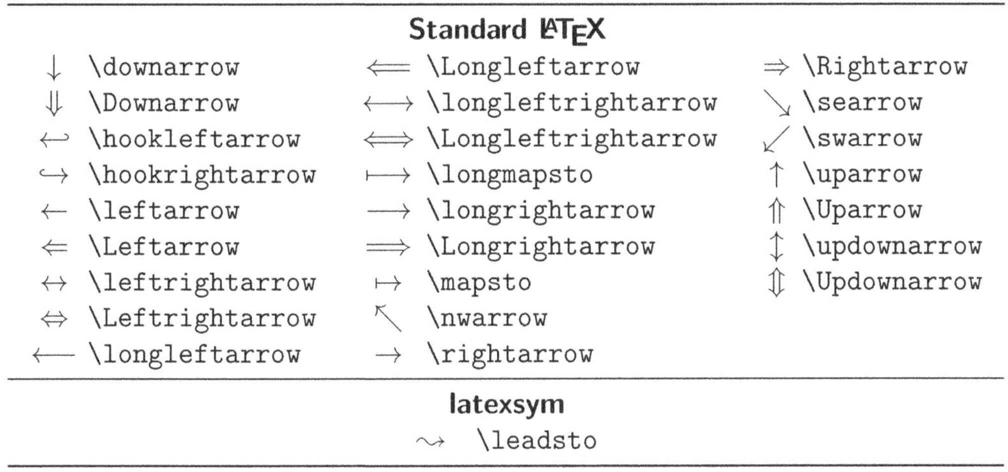

Standard LaTeX		
$\downarrow$ \downarrow	$\Longleftarrow$ \Longleftarrow	$\Rightarrow$ \Rightarrow
$\Downarrow$ \Downarrow	$\longleftrightarrow$ \longleftrightarrow	$\searrow$ \searrow
$\hookleftarrow$ \hookleftarrow	$\Longleftrightarrow$ \Longleftrightarrow	$\swarrow$ \swarrow
$\hookrightarrow$ \hookrightarrow	$\longmapsto$ \longmapsto	$\uparrow$ \uparrow
$\leftarrow$ \leftarrow	$\longrightarrow$ \longrightarrow	$\Uparrow$ \Uparrow
$\Leftarrow$ \Leftarrow	$\Longrightarrow$ \Longrightarrow	$\updownarrow$ \updownarrow
$\leftrightarrow$ \leftrightarrow	$\mapsto$ \mapsto	$\Updownarrow$ \Updownarrow
$\Leftrightarrow$ \Leftrightarrow	$\nwarrow$ \nwarrow	
$\longleftarrow$ \longleftarrow	$\rightarrow$ \rightarrow	
latexsym		
$\leadsto$ \leadsto		

Figure 9.23 – Arrows

Arrows are used in many contexts: for implications, mappings between sets, limits, sequences, vector notation, and various descriptive expressions.

Harpoons

LaTeX provides a special kind of arrow symbol called **harpoons**, which look like one-sided arrowheads:

$\leftharpoondown$ \leftharpoondown	$\rightharpoondown$ \rightharpoondown	$\rightleftharpoons$ \rightleftharpoons
$\leftharpoonup$ \leftharpoonup	$\rightharpoonup$ \rightharpoonup	

Figure 9.24 – Harpoons

Harpoons are used, for example, in chemical reaction formulas.

Symbols derived from letters

Mathematics uses various symbols that resemble stylized letters:

Standard LaTeX				
$\bot$ \bot	$\forall$ \forall	$\imath$ \imath	$\ni$ \ni	$\top$ \top
ℓ \ell	$\hbar$ \hbar	$\in$ \in	∂ \partial	$\wp$ \wp
$\exists$ \exists	$\Im$ \Im	$\jmath$ \jmath	$\Re$ \Re	
latexsym				
$\mho$ \mho				

Figure 9.25 – Symbols derived from letters

Mathematicians often use `\in`, `\forall`, and `\exists` in statements.

Miscellaneous symbols

Here are some LaTeX symbols that do not fit into the categories mentioned previously:

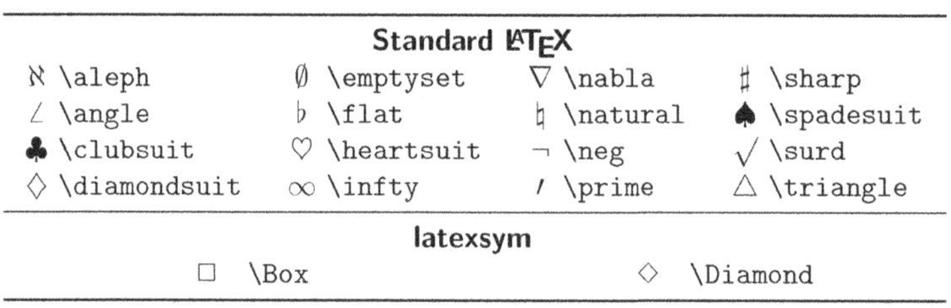

Standard LaTeX			
$\aleph$ \aleph	$\emptyset$ \emptyset	∇ \nabla	$\sharp$ \sharp
$\angle$ \angle	$\flat$ \flat	$\natural$ \natural	$\spadesuit$ \spadesuit
$\clubsuit$ \clubsuit	$\heartsuit$ \heartsuit	$\neg$ \neg	$\surd$ \surd
$\diamondsuit$ \diamondsuit	∞ \infty	$\prime$ \prime	$\triangle$ \triangle
latexsym			
$\Box$ \Box	$\Diamond$ \Diamond		

Figure 9.26 – Additional LaTeX symbols

The Comprehensive LaTeX Symbol List lists over 20,000 symbols sorted into categories in one single PDF document. If you need to look up a symbol, this document is the best place to start. With TeX Live, you can open it at the command prompt as follows:

```
texdoc symbols
```

Or you can visit `https://texdoc.org/pkg/symbols` to browse it.

Another fascinating approach is handwritten symbol recognition. You draw a symbol with the mouse or your finger on a touchscreen, and the software tries to recognize it and tells you its code. Let's have a quick look:

1. Visit `https://detexify.kirelabs.org`.
2. Draw the symbol in the white box. It doesn't matter if it's a shaky mouse sketch, like this:

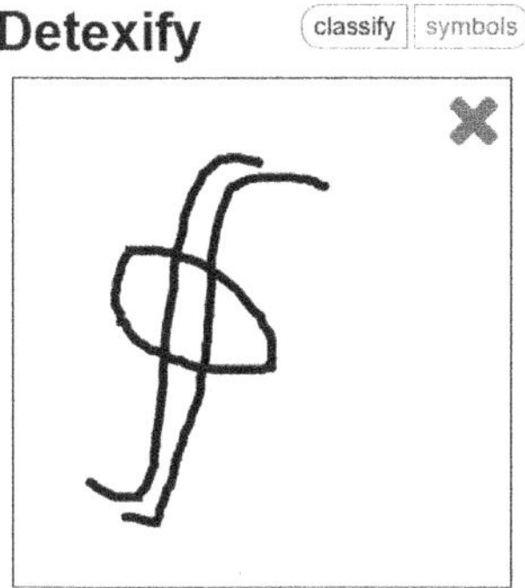

Figure 9.27 – Handwritten symbol

3. After a moment, the tool shows suggestions for symbols along with the corresponding LaTeX commands:

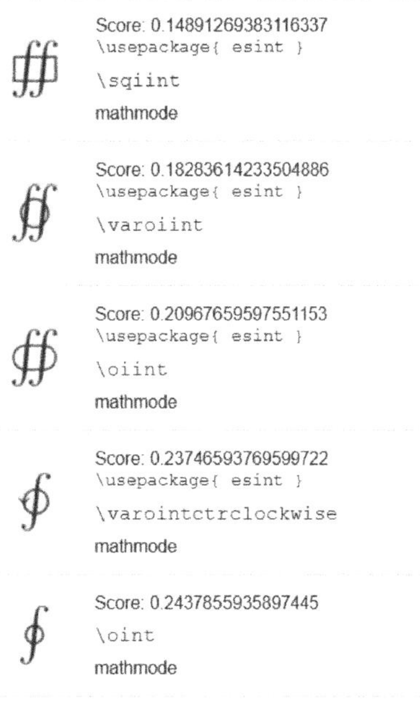

Figure 9.28 – Symbol and code suggestions

Detexify also provides a name-based search. Click the **Symbols** button at the top, enter a phrase into the filter, and Detexify will show symbols and commands that match your query.

Writing units

When writing **units** in text, they should not look like variables. For instance, m for meters should not look exactly like an m variable, and s may stand for seconds, but not for an s variable. A typographical convention is to use an upright font shape for units, while variables are written in italics. It's also common to place a thin space between the value and the unit. So, for 10 meters, you may write `10\,\mathrm{m}`. In physics, chemistry, and engineering, we have much more complex units, such as **meters per second squared** (**m/s²**) for acceleration. The `siunitx` package supports correct and consistent typesetting of such units. It requires reading some documentation before you can use it, but it's worth the effort. Run `texdoc siunitx` at the command prompt or visit `https://texdoc.org/pkg/siunitx`.

Variable-sized operators

For sums, products, integrals, and set operations, LaTeX provides operator symbols that adjust their size automatically: they appear bigger in **display style** and smaller in **text style**.

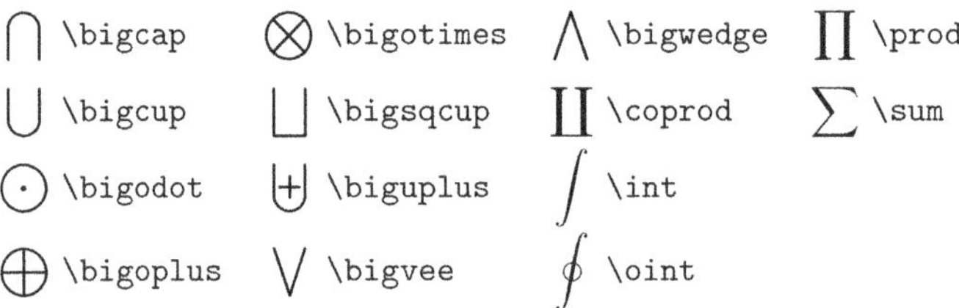

Figure 9.29 – Variable-sized operators

Here's an equation in text style:

```
\(
  \int_a^b \! f(x) \, dx = \lim_{\Delta x \rightarrow 0}
  \sum_{i=1}^{n} f(x_i) \,\Delta x_i
\)
```

This code gives us the following:

$$\textstyle \int_a^b \! f(x) \, dx = \lim_{\Delta x \rightarrow 0} \sum_{i=1}^{n} f(x_i) \,\Delta x_i$$

Figure 9.30 – Inline text style equation

And here's the same equation in displayed style:

```
\[
  \int_a^b \! f(x) \, dx = \lim_{\Delta x \rightarrow 0}
  \sum_{i=1}^{n} f(x_i) \,\Delta x_i
\]
```

This time, we get the following:

$$\int_a^b \! f(x) \, dx = \lim_{\Delta x \rightarrow 0} \sum_{i=1}^{n} f(x_i) \,\Delta x_i$$

Figure 9.31 – Displayed style equation

As you can see, the operator symbols are noticeably larger in displayed style.

Variable-sized delimiters

Delimiters such as **parentheses**, **brackets**, and **braces** can change size depending on the expression they enclose. The following are such LaTeX delimiters:

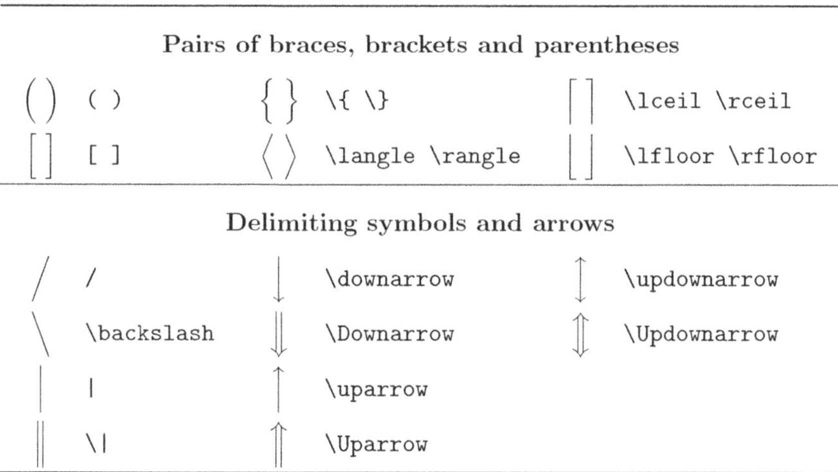

Pairs of braces, brackets and parentheses					
$\big(\big)$	( )	$\big\{ \big\}$	\{ \}	$\big\lceil \big\rceil$	\lceil \rceil
$\big[\big]$	[ ]	$\big\langle \big\rangle$	\langle \rangle	$\big\lfloor \big\rfloor$	\lfloor \rfloor
Delimiting symbols and arrows					
$\big/$	/	$\big\downarrow$	\downarrow	$\big\updownarrow$	\updownarrow
$\big\backslash$	\backslash	$\big\Downarrow$	\Downarrow	$\big\Updownarrow$	\Updownarrow
$\big\vert$	|	$\big\uparrow$	\uparrow		
$\big\Vert$	\\|	$\big\Uparrow$	\Uparrow		

Figure 9.32 – Variable-sized delimiters

LaTeX provides special commands to adjust them automatically. If you place a `\left` or `\right` command right before such a delimiter, LaTeX automatically matches its size to the height of the enclosed expression. We have to use these size macros in pairs. To match a pair, if you don't want a second delimiter, use `\left.` or `\right.` to get an invisible delimiter on one side.

Automatically adjusting delimiters is especially useful for larger structures, such as matrices, so let's look at those next.

Building math structures

Variables and constants are straightforward, but mathematical writing often involves more complex objects, such as binomial coefficients, vectors, and matrices. In this section, we'll look at how to typeset such structures.

Let's begin with simple arrays.

Creating arrays

To arrange mathematical expressions inside a larger expression, you can use the `array` environment. It works much like a `tabular` environment, but it's for math mode, and all its entries are written in math mode as well.

For example, you can place an array inside variable-sized parentheses:

```
\[
  A = \left(
    \begin{array}{cc}
      a_{11} & a_{12} \\
      a_{21} & a_{22}
    \end{array}
  \right)
\]
```

This produces a matrix:

$$A = \left(\begin{array}{cc} a_{11} & a_{12} \\ a_{21} & a_{22} \end{array} \right)$$

Figure 9.33 – A simple array

There are also dedicated commands for matrices, which we'll look at next.

Typesetting matrices

The `amsmath` package provides several matrix environments. A standard matrix can be produced with the `pmatrix` environment:

```
\documentclass{article}
```

```
\usepackage{amsmath}
\begin{document}
\[
  A = \begin{pmatrix}
    a_{11} & a_{12} \\
    a_{21} & a_{22}
  \end{pmatrix}
\]
\end{document}
```

This yields the following:

$$A = \begin{pmatrix} a_{11} & a_{12} \\ a_{21} & a_{22} \end{pmatrix}$$

Figure 9.34 – A simple matrix

You may notice that the parentheses sit closer to the matrix entries than in the array example in the previous section. This tighter spacing is part of the `amsmath` style.

Here are the `amsmath` matrix environments and their delimiters:

- `matrix`: No delimiters
- `pmatrix`: Parentheses, ( )
- `bmatrix`: Square brackets, [ ]
- `Bmatrix`: Braces, { }
- `vmatrix`: | |
- `Vmatrix`: || ||
- `smallmatrix`: More compact, no delimiters; you can add them manually, if needed

The compact `smallmatrix` environment is useful for matrices that appear within regular text.

Writing binomial coefficients

You could write **binomial coefficients** using an array together with delimiters, but the `amsmath` package provides a shorter and cleaner command: `\binom` for binomial coefficients. Try this:

```
\binom{n}{k} = \frac{n!}{k!(n-k)!}
```

The result is as follows:

$$\binom{n}{k} = \frac{n!}{k!(n-k)!}$$

Figure 9.35 – A binomial coefficient in an equation

This syntax is much simpler than using an array or a matrix for such a small expression.

Underlining and overlining

The `\overline` command puts a line above its argument:

```
\overline{\Omega}
```

This produces the following:

Figure 9.36 – An overlined omega symbol

The corresponding command for placing a line below is `\underline`.

You can also use braces instead of lines. The `\underbrace` and `\overbrace` commands create braces below or above an expression:

```
N = \underbrace{1 + 1 + \cdots + 1}_n
```

This gives the following:

$$N = \underbrace{1 + 1 + \cdots + 1}_n$$

Figure 9.37 – An underbrace below an expression

A subscript placed on an underbrace is written below it, and a superscript placed on an overbrace appears above.

Setting accents

In *Chapter 2, Formatting Text and Creating Macros,* we've already seen how to write accents in text mode. For math mode, we need different commands. We can apply them to any letter. Here's the list of math accents shown using the lowercase letter *a* as an example:

$\acute{a}$ \acute{a}	$\check{a}$ \check{a}	$\grave{a}$ \grave{a}	$\tilde{a}$ \tilde{a}
$\bar{a}$ \bar{a}	$\ddot{a}$ \ddot{a}	$\hat{a}$ \hat{a}	$\vec{a}$ \vec{a}
$\breve{a}$ \breve{a}	$\dot{a}$ \dot{a}	$\mathring{a}$ \mathring{a}	
Extensible			
$\widehat{abc}$ \widehat{abc}		$\widetilde{abc}$ \widetilde{abc}	

Figure 9.38 – Various math accents

Extensible accents, also called **wide accents**, automatically adjust to the width of their arguments.

Putting a symbol above or below another one

Beyond the `array` environment, `amsmath` provides commands for stacking expressions directly:

- `\underset{expression below}{expression}` places one expression below another, using subscript-sized text
- `\overset{expression above}{expression}` puts an expression above another, using superscript-sized text

This is an example of how we can use these commands:

```
\underset{\circ}{\cap} \neq \overset{\circ}{\cup}
```

This gives us the following:

$$\underset{\circ}{\cap} \neq \overset{\circ}{\cup}$$

Figure 9.39 – Putting a symbol above or below another one

Another handy command is `\stackrel{expression above}{relation}`. See the following formula, for example:

```
X \stackrel{\text{def}}{=} 0
```

This results in the following:

$$X \stackrel{\text{def}}{=} 0$$

Figure 9.40 – Text above a relation symbol

`\stackrel` puts an expression above a relation symbol.

Writing theorems and definitions

LaTeX provides dedicated environments for **theorems**, **definitions**, and similar statements. Returning to our first example in this chapter, we could use the `\newtheorem` command to define a `Theorem` environment, `thm`, as follows:

```
\newtheorem{thm}{Theorem}
```

Next, we can declare a `Definition` environment. Let's call it `dfn` here:

```
\newtheorem{dfn}[thm]{Definition}
```

We can use an optional argument referring to an existing environment (in this case, `thm`). This makes the new environment use the same counter as the already existing environment. In our case here, this means that after *Theorem 1* follows *Definition 2*.

We can use these environments as follows:

```
\begin{dfn}
  A quadratic equation is an equation of the form
  \begin{equation}
    \label{quad}
    ax^2 + bx + c = 0,
  \end{equation}
  where \( a, b \) and \( c \) are constants and \( a \neq 0 \).
\end{dfn}
\begin{thm}
  A quadratic equation (\ref{quad}) has two solutions for the
  Variable \( x \):
  \begin{equation}
    \label{root}
    x_{1,2} = \frac{-b \pm \sqrt{b^2-4ac}}{2a}.
  \end{equation}
\end{thm}
```

Take a look at the output:

Definition 1 *A quadratic equation is an equation of the form*

$$ax^2 + bx + c = 0, \tag{1}$$

where a, b and c are constants and $a \neq 0$.

Theorem 2 *A quadratic equation (1) has two solutions for the Variable x:*

$$x_{1,2} = \frac{-b \pm \sqrt{b^2 - 4ac}}{2a}. \tag{2}$$

Figure 9.41 – A definition and a theorem

In the output, such environments are numbered and labeled `Definition` and `Theorem`, respectively. In *Chapter 11, Developing Large Documents*, we will use this approach to create a complete document containing definitions, theorems, and lemmas.

There are two special packages offering much more flexibility:

- `amsthm` provides several styles, allows fine-grained customization, and includes a proof environment
- `ntheorem` offers similar features but handles traditional *quod erat demonstrandum* (*what was to be shown*) end marks for proofs in a better way

If you would like to use such environments, look at their documentation and compare the relevant features to decide which package is best for you. As usual, execute `texdoc amsthm` and `texdoc ntheorem` at the command prompt, or visit `https://texdoc.org/pkg/amsthm` and `https://texdoc.org/pkg/ntheorem`.

Choose one of those closely related packages; don't load both.

Extended math tools

Take a look at all the mathematics typesetting options in `amsmath` by entering `texdoc amsmath` at the command line or visiting `https://texdoc.org/pkg/amsmath`.

The `mathtools` package extends `amsmath` and adds many features. If you are looking for something that standard LaTeX or `amsmath` doesn't offer, `mathtools` is the next place to check. Here are some of its features:

- Tools for fine-tuning mathematical typesetting, such as more compact superscript styles
- Vertical alignment of limits for consecutive operators
- Adjusting the width of operators
- Improved control over tags, including modifying their appearance and only showing tags for equations that have been referenced
- Extensible symbols – more arrows with automatic width adjustment, as well as extensible brackets and braces to be set under or over expressions
- New math environments for more flexible matrices, cases, improved multi-line formulas, and arrows between aligned expressions
- Tighter spacing for shorter intertext
- Declaration of paired delimiters
- Additional symbols, such as a vertically centered colon, along with combinations of relation symbols with colons and shortcuts for auto-sized parentheses
- Techniques for spreading lines in multi-line formulas, setting left subscripts and superscripts, typesetting math within italicized text, and creating multi-line fractions

Have a look at the documentation of this valuable package and find out which commands can be applied to achieve the styles and alignments listed here. Open it by running `texdoc mathtools` at the command line or go to `https://texdoc.org/pkg/mathtools`.

In *Chapter 10*, *Writing Advanced Mathematics*, in *LaTeX Cookbook*, you can find a lot of examples showing enhancements with the `mathtools` package. Visit `https://latex-cookbook.net/tag/mathematics/` to see and run examples online that include fine-tuning math formulas, automatic line breaking in equations, plotting functions, and drawing diagrams and geometry pictures.

Summary

You can now write complex math formulas, and you've got the essential tools for writing scientific texts. We worked with the `amsmath` package, which provides us with many features tailored to traditional mathematical typesetting.

You can find further code examples in *Chapter 10*, *Writing Advanced Mathematics*, of *LaTeX Cookbook*, with compilable code on the book's website at `https://latex-cookbook.net/chapter-10`.

At this point, you can now fine-tune math expressions, align and number equations, and work with a wide range of mathematical symbols from various symbol fonts. In the next chapter, we'll look at fonts in general.

10

Using Fonts

With mathematical typesetting covered, we can now look at the fonts that define a document's look. This chapter introduces LaTeX's font options and shows how text and math fonts work together. The base font you choose strongly shapes the overall appearance. You might select a clear, easy-to-read font for a lengthy document or a decorative, calligraphic font for a greeting card. Your job application letter could benefit from a very clean professional font. In contrast, a mathematical article requires fonts with a wide range of symbols and a text font that complements them.

Until now, we have focused on the logical properties of fonts. Although we've always used the LaTeX standard font, we switched from Roman to sans-serif and typewriter styles, and learned how to make text bold, italic, and slanted back in *Chapter 2, Formatting Text and Creating Macros*. However, we've not yet moved beyond the standard font set.

In this chapter, we'll look at the following topics:

- Using comprehensive font bundles
- Choosing specific font families
- Writing with TrueType and OpenType system fonts

While we'll examine the text's appearance, we'll also look at how math formulas look together with the text font.

As this book is printed and electronically distributed with bitmap images, you will not see the original LaTeX quality in the font samples in this chapter. Visit `https://latexguide.org/chapter-10` to see the fonts in original LaTeX and PDF quality. You can see the font examples in a larger size to easily highlight the fine details and differences.

Let's begin with an example. It will serve as the basis for our work with fonts throughout this chapter.

Using comprehensive font bundles

We begin with the most extensive font bundles. To test fonts, it helps to use a **pangram**. This word comes from the Greek *pan gramma*, meaning *every letter*. It stands for a sentence that uses every letter of the alphabet. Pangrams are ideal for showing how a font handles its full character set.

We will print a famous pangram phrase using the **Latin Modern** font family. Latin Modern is very similar to LaTeX's default font, **Computer Modern**. However, Latin Modern includes many additional characters, especially accented ones. Because of this broader coverage and its high quality, it's often considered the natural successor to the standard font. Let's see how it looks in various font families and shapes, together with a math formula:

1. Start a new document:

   ```
   \documentclass{article}
   ```

2. Create a macro for the pangram and include additional numerals. It shall take one argument, which will be the font family or shape selection command. We will add a paragraph break at the end, as shown:

   ```
   \newcommand{\pangram}[1]{{#1 The quick brown fox
   jumps over the lazy dog. 1234567890\par}}
   ```

3. Load the `fontenc` package and choose the `T1` font encoding:

   ```
   \usepackage[T1]{fontenc}
   ```

4. Load the `lmodern` package to get the Latin Modern font:

   ```
   \usepackage{lmodern}
   ```

5. Begin the document and choose a `large` font size so the details are easy to see:

   ```
   \begin{document}
   \large
   ```

6. Now, we use our `\pangram` macro several times with different font settings:

   ```
   \pangram{\rmfamily}
   \pangram{\sffamily}
   \pangram{\ttfamily}
   ```

```
\pangram{\itshape}
\pangram{\slshape}
```

7. Add a math font example; use the code that we wrote for *Figure 9.29* in *Chapter 9, Writing Math Formulas*:

```
\[
  \int_a^b \! f(x) \, dx = \lim_{\Delta x \rightarrow 0}
  \sum_{i=1}^{n} f(x_i) \,\Delta x_i
\]
\end{document}
```

8. Compile and look at the font examples:

The quick brown fox jumps over the lazy dog. 1234567890
The quick brown fox jumps over the lazy dog. 1234567890
The quick brown fox jumps over the lazy dog.
1234567890
The quick brown fox jumps over the lazy dog. 1234567890
The quick brown fox jumps over the lazy dog. 1234567890

$$\int_a^b \! f(x) \, dx = \lim_{\Delta x \rightarrow 0} \sum_{i=1}^{n} f(x_i) \,\Delta x_i$$

Figure 10.1 – Latin Modern font examples

In our `\pangram` macro definition, we had another pair of curly braces. In our `{{ ... }}` argument, the outer curly braces hold the argument to `\newcommand`, and the inner curly braces limit our commands to restrict the effect of the font change.

Remember, the `\pangram` macro is just a small demonstration macro for us, so we won't have to repeat the demo sentence for each font family. In your everyday documents, load the font package and write your text. If needed, you can switch between font families such as `\sffamily`, `\ttfamily`, or `\rmfamily`, just like in *Chapter 2, Formatting Text and Creating Macros*.

In *step 3*, we selected a **font encoding**. Technically, encodings are mappings of character codes to symbols in a font. For English and most Western European languages, the `T1` font encoding is recommended. It's also known as **Cork encoding** because it was developed in the city of Cork, Ireland, during a TeX user group conference.

The default LaTeX font encoding is called `OT1`. Compared to `OT1`, the `T1` encoding uses larger encoding tables that handle accented characters much better.

For example, with the default historic `OT1` encoding, the accented ö character is built from the o **glyph** and separate dots, to be printed in the PDF file. With `T1`, an ö is a single glyph of the current font, so LaTeX can also properly apply hyphenation rules to words containing accented characters. The search feature of a PDF reader also works with those characters, and copying and pasting from the PDF file preserves them. With the default `OT1` encoding, copying and pasting the ö character would result in dots and an o character.

If you notice a drop in default font quality with `T1` encoding, your installation may be missing fonts. In that case, install the `cm-super` package using the package manager, or switch to one of the fonts with `T1` support described in the next sections.

What we have discussed so far is **output encoding**. You may also encounter the term **input encoding**. Modern operating systems and editors support **UTF-8 (Unicode)**, an industry-standard text encoding that extends the aged **ASCII** code, which can handle only 128 characters. LaTeX supports UTF-8 natively, so we won't need to do anything. So, if you encounter the `inputenc` package in older books or code on the internet, you may omit it. A good practice is to use UTF-8 with any editor.

Now, we will take a look at some recommended fonts with examples. All of them support `T1` encoding, so before you load a font, use the following command:

```
\usepackage[T1]{fontenc}
```

In the following sections, we will explore the different fonts.

Latin Modern – a replacement for the standard font

Latin Modern was designed to look like the default LaTeX font, but the encoding has been improved, and it has also received additional refinements. Latin Modern contains many **diacritic characters** as single glyphs, whereas Computer Modern builds such accented characters from plain letters and accents.

Latin Modern has 72 text fonts and 20 math fonts under the hood, supporting all font families, shapes, and weights.

In *Figure 10.1*, we saw what it looks like.

Kp-Fonts – another extensive set of fonts

The **Kp-Fonts** collection from the **Johannes Kepler project** provides serif, sans-serif, and mono-spaced fonts, as well as mathematical symbol fonts in several shapes and weights. It even includes bold-extended and combinations such as slanted serif small caps.

Just load the package to use those fonts:

```
\usepackage{kpfonts}
```

The previous example will change to the following:

The quick brown fox jumps over the lazy dog. 1234567890
The quick brown fox jumps over the lazy dog. 1234567890
The quick brown fox jumps over the lazy dog.
1234567890
The quick brown fox jumps over the lazy dog. 1234567890
The quick brown fox jumps over the lazy dog. 1234567890

$$\int_a^b f(x)\,dx = \lim_{\Delta x \to 0} \sum_{i=1}^{n} f(x_i)\,\Delta x_i$$

Figure 10.2 – Kepler font examples

Kp-Fonts also provides **light versions** with the same font metrics. The light versions might look nice in print but may appear a bit too thin for screen reading.

To switch to the light versions, load the package with the `light` option:

```
\usepackage[light]{kpfonts}
```

The look will be different now:

The quick brown fox jumps over the lazy dog. 1234567890
The quick brown fox jumps over the lazy dog. 1234567890
The quick brown fox jumps over the lazy dog.
1234567890
The quick brown fox jumps over the lazy dog. 1234567890
The quick brown fox jumps over the lazy dog. 1234567890

$$\int_a^b f(x)\,dx = \lim_{\Delta x \to 0} \sum_{i=1}^{n} f(x_i)\,\Delta x_i$$

Figure 10.3 – Kepler font as a light version

Now, let's look at more specialized font packages that provide a single style.

Choosing specific font families

We'll now look at several TeX fonts that each have their own character and style. As before, we'll use our `\pangram` macro from the previous section combined with the corresponding font-family command for testing.

Serif fonts

A small line or **stroke** attached to the ends of a larger stroke in a letter or symbol is called a **serif**. A font using such serifs is called a **serif font**, or a **serif typeface**.

The default serif font is called **Computer Modern Roman**. Latin Modern provides a very similar look, and you already know the Kp-Fonts serif font. Other packages specialize in serif fonts, and we will now look at some of them.

Times Roman

The `newtx` package defines a **Times**-style text font along with a matching math font. The package is split into two parts so that the text and math components can be loaded independently, which is helpful when you want to combine Times with a different math font. That's why we load it in this way:

```
\usepackage{newtxtext}
\usepackage{newtxmath}
```

With `\pangram{\rmfamily}` and our math formula, we get the following:

The quick brown fox jumps over the lazy dog. 1234567890

$$\int_a^b f(x)\,dx = \lim_{\Delta x \to 0} \sum_{i=1}^{n} f(x_i)\,\Delta x_i$$

Figure 10.4 – A Times Roman font

As you can see, Times is a very narrow font that works well in multi-column text, as in newspapers, but is not well-suited to single-column text. Wide text lines with too many characters are not ideal.

Palatino

The `newpx` package provides a **Palatino**-style text font and a matching math font. This package is also split into two parts for independent use, so we load them separately:

```
\usepackage{newpxtext}
\usepackage{newpxmath}
```

This gives us the following:

The quick brown fox jumps over the lazy dog. 1234567890

$$\int_a^b f(x)\,dx = \lim_{\Delta x \to 0} \sum_{i=1}^{n} f(x_i)\,\Delta x_i$$

Figure 10.5 – A Palatino font

We can see that Palatino is noticeably wider than Times, giving it a more open and relaxed appearance.

Charter

Charter is similar to the default Computer Modern font but has a slightly heavier appearance. We load it like this:

```
\usepackage{charter}
```

For proper math support, load the `mathdesign` package with the `charter` option instead of loading `charter` directly:

```
\usepackage[charter]{mathdesign}
```

This produces the following:

The quick brown fox jumps over the lazy dog. 1234567890

$$\int_a^b f(x)\,dx = \lim_{\Delta x \to 0} \sum_{i=1}^{n} f(x_i)\,\Delta x_i$$

Figure 10.6 – Charter and mathdesign fonts

Besides `Charter`, the `mathdesign` package can also load the Utopia font:

```
\usepackage[utopia]{mathdesign}
```

And this loads the Garamond font:

```
\usepackage[garamond]{mathdesign}
```

New Century Schoolbook

The `newcent` package provides the easy-to-read serif typeface **New Century Schoolbook**:

```
\usepackage{newcent}
```

To pair it with a suitable math font, you can load the **Fourier** math fonts:

```
\usepackage{fouriernc}
```

Here they are together:

The quick brown fox jumps over the lazy dog. 1234567890

$$\int_a^b f(x)dx = \lim_{\Delta x \to 0} \sum_{i=1}^n f(x_i)\Delta x_i$$

Figure 10.7 – New Century Schoolbook and Fourier fonts

`nc` in the `fouriernc` package stands for *New Century* because it's designed as a companion.

Concrete Roman

The **Concrete Roman** font may not look ideal on screen, but it delivers excellent print quality. Just load the `concrete` package:

```
\usepackage{concrete}
```

A matching math companion is available through the `concmath` package:

```
\usepackage{concmath}
```

Together, they produce the following result:

The quick brown fox jumps over the lazy dog. 1234567890

$$\int_a^b f(x)\,dx = \lim_{\Delta x \to 0} \sum_{i=1}^n f(x_i)\,\Delta x_i$$

Figure 10.8 – Concrete Roman font with math support

With its upright integral sign and serifless summation symbol, Concrete Roman has a distinctive appearance.

Bookman

Bookman is an old-style serif font, provided through the `bookman` package, loaded by the following command:

```
\usepackage{bookman}
```

The **Kerkis** font is a `bookman` extension with math support; you can load this instead:

```
\usepackage{kmath}
\usepackage{kerkis}
```

We get the following:

The quick brown fox jumps over the lazy dog. 1234567890

$$\int_a^b f(x)\,dx = \lim_{\Delta x \to 0} \sum_{i=1}^{n} f(x_i)\,\Delta x_i$$

Figure 10.9 – Kerkis, aka Bookman, with math support

An even more enhanced Bookman-inspired font is available under the name **TeX Gyre Bonum**. This one, though, especially with math support, is better used as an OpenType font. In the last section of this chapter, we will deal with this.

Fonts with the same or similar design often appear under different names. That's often for legal reasons, since the font names can be protected even when the design itself is free to use.

Sans-serif fonts

Sans-serif fonts are simply fonts where no serifs are used. They tend to look clearer, making them a popular choice for slide presentations and other screen-based content.

Because sans-serif fonts don't appear as heavy in bold, they work well for headings. However, many believe that longer passages of text are much more readable with traditional serifs. That's the reason why **KOMA-Script** classes use a serif font in the document body text and a sans-serif font for the headings by default.

If required, the main body font could be rendered sans-serif by means of this command:

```
\renewcommand{\familydefault}{\sfdefault}
```

We already know that Latin Modern and Kp-Fonts provide sans-serif variants. Let's now take another look at some dedicated sans-serif fonts.

Arev

Arev is a sans-serif font designed for slide presentations. The name is *Vera* spelled backward, as it extends the **Vera Sans** font, itself based on the **Frutiger** font. Arev also includes matching math support. Load it like this:

```
\usepackage{arev}
```

Text and math become the following:

The quick brown fox jumps over the lazy dog. 1234567890

$$\int_a^b f(x)\,dx = \lim_{\Delta x \to 0} \sum_{i=1}^{n} f(x_i)\,\Delta x_i$$

Figure 10.10 – Arev, a Frutiger-like font

Note that integral and summation signs still retain serifs, as this is very common.

Computer Modern Bright

Computer Modern Bright (**CM Bright**) is derived from Computer Modern Sans Serif to produce a lighter variant. The `cmbright` package provides this font together with a light typewriter font and a sans-serif math font. Load it as follows:

```
\usepackage{cmbright}
```

The output of our sample code will be like this:

The quick brown fox jumps over the lazy dog. 1234567890

$$\int_a^b f(x)\,dx = \lim_{\Delta x \to 0} \sum_{i=1}^{n} f(x_i)\,\Delta x_i$$

Figure 10.11 – The CM Bright font

Compared with other sans-serif fonts, it's less bold and more understated. Because of its light weight, it pairs best with text fonts of similar lightness rather than with heavier serif fonts.

Kurier

Many sans-serif fonts look similar at first glance, but their differences set them apart. Look at the **Kurier** font in *Figure 10.12*, for example; you can notice differences, especially in the shape of the letter *g* and the math symbols. Load it as follows:

```
\usepackage{kurier}
```

We can enable math support by using the `math` option:

```
\usepackage[math]{kurier}
```

Compiling our example code gives us the following:

The quick brown fox jumps over the lazy dog. 1234567890

$$\int_a^b f(x)\,dx = \lim_{\Delta x \to 0} \sum_{i=1}^{n} f(x_i)\,\Delta x_i$$

Figure 10.12 – The Kurier font

In this font family, even the integral and summation symbols appear without serifs.

Helvetica

The classic sans-serif font **Helvetica** is simple, clean, and widely recognized. You probably know its Microsoft descendant, **Arial**. Load the font this way:

```
\usepackage{helvet}
```

Use the `scaled` option if the font appears too large, especially when next to a serif font. For example, to scale it a bit down, you can write the following:

```
\usepackage[scaled=0.95]{helvet}
```

Helvetica doesn't include direct math support, but the `sfmath` package can help:

```
\usepackage{sfmath}
```

When you load the `sfmath` package in your document preamble, LaTeX will also use the current sans-serif text font within math formulas. Place it after other font packages, so it has a chance to detect the font. Further explanations, examples, options, and an alternative approach using the `sansmath` package can be found in *Chapter 3, Adjusting Fonts*, of the *LaTeX Cookbook*, with the source code at `https://latex-cookbook.net/chapter-03`.

Loading `helvet` and `sfmath` as in this section produces the following output:

The quick brown fox jumps over the lazy dog. 1234567890

$$\int_a^b f(x)\,dx = \lim_{\Delta x \to 0} \sum_{i=1}^{n} f(x_i)\,\Delta x_i$$

Figure 10.13 – Helvetica example

You can find further information on the `sfmath` author's home page at `https://dtrx.de/od/tex/sfmath.html`.

Typewriter fonts

Typewriter fonts, also known as **monospaced** fonts, are commonly used for source code, as in this book. Let's look at three excellent variants.

Courier

Courier is a very wide-running typewriter font. Load it with this:

```
\usepackage{courier}
```

Then, with `\ttfamily` or `\texttt`, we will get the following:

`The quick brown fox jumps over the lazy dog. 1234567890`

Figure 10.14 – The Courier font

If that appears too big compared to your main document font, you can load the `couriers` package instead (note the *s* for *scaled*) with a `scaled` option, such as here:

```
\usepackage[scaled=0.95]{couriers}
```

This reduces the Courier font to 95% of its original size.

Inconsolata

Inconsolata is a well-designed monospaced font made specifically for source code listings. It's more compact and easier to read than Courier. Load it as follows:

```
\usepackage{inconsolata}
```

The output proves that monospaced fonts can be elegant:

The quick brown fox jumps over the lazy dog. 1234567890

Figure 10.15 – The Inconsolata font

Unlike Courier, it's sans-serif. It also supports a `scaling` option if you need to adjust its size.

Bera Mono

Bera Mono is another sans-serif typewriter font. Load it like this:

```
\usepackage{beramono}
```

This is how it looks:

The quick brown fox jumps over the lazy dog. 1234567890

Figure 10.16 – The Bera Mono font

Here, you can also specify a `scaled` option.

Calligraphic fonts

Calligraphic fonts are **script typefaces** with fluid strokes similar to handwriting. They work nicely for invitations, decorative headings, or any text that needs a fancy touch. Let's pick two beautiful handwriting fonts for a closer look.

Calligra

We load the font in the usual way:

```
\usepackage{calligra}
```

To switch to the font, we can use the `\calligra` command in the text. As we know, local switching commands are valid until a surrounding environment or group, { ... }, ends. It also works with our `\pangram` macro, like this:

```
\pangram{\calligra}
```

This prints the following:

The quick brown fox jumps over the lazy dog. 1234567890

Figure 10.17 – The Calligra handwriting font

Capital letters look particularly lively and playful.

Miama Nueva

The parts of letters that extend above the normal letter height are called **ascenders**, while those that drop below the baseline are called **descenders**. **Miama Nueva** is an elegant handwriting font with especially graceful ascenders and descenders. Load it as usual:

```
\usepackage{miama}
```

Then, the `\fmmfamily` command switches to that font. Again, use it within a group, `{ ... }`, or an environment if you want to limit the font choice to just a piece of text. We can use our `\pangram` macro again:

```
\pangram{\fmmfamily}
```

The writing is a pleasure to read:

The quick brown fox jumps over the lazy dog. 1234567890

Figure 10.18 – The Miama Nueva handwriting font

Miama Nueva is very charming, for example, on invitation cards for weddings.

Probably the best place to browse LaTeX fonts is **The LaTeX Font Catalogue**. You can visit it online at `https://www.tug.org/FontCatalogue`. The site aims to present all freely available fonts for LaTeX. It's based on TeX Live. It shows visual samples, the required code, and further useful information. Pick a category, browse the previews, and click a font to see some examples, usage notes, and the required code.

You can also use fonts that LaTeX doesn't support directly, and that's what we'll look at next.

Writing with TrueType and OpenType system fonts

LaTeX can also use many thousands of fonts that were never even prepared for LaTeX. That includes operating system fonts, **TrueType** fonts, and modern **OpenType** fonts.

Let's try this with fonts available on a **Microsoft Windows** computer. For macOS users, we will look at an example afterward.

Selecting the main font

We can either open **Settings** | **Fonts** through the Windows **Start** menu or look in the `C:\Windows\Fonts` folder to see the installed fonts. The **Segoe UI** font appears available with several names, so let's choose **Segoe UI Semilight**. Let's see how easy it is to use:

1. Start a new document:

   ```
   \documentclass{article}
   ```

2. Load the `fontspec` package, which provides the font selection commands:

   ```
   \usepackage{fontspec}
   ```

3. Set the main font:

   ```
   \setmainfont{Segoe UI Semilight}
   ```

4. Write the document body with some large text:

   ```
   \begin{document}
   \large
   The quick brown fox jumps over the lazy dog. 1234567890
   \end{document}
   ```

5. This time, choose either **LuaLaTeX** or **XeLaTeX** as the compiling engine. In TeXworks, this is a drop-down list right next to the **Typeset** button, as shown here:

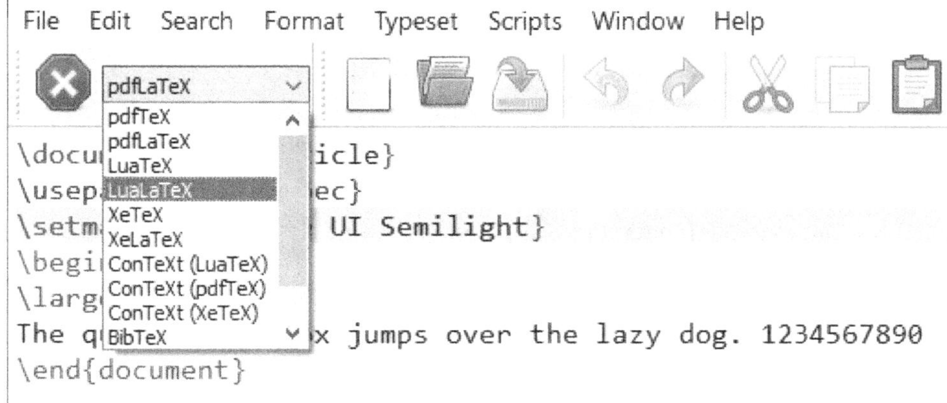

Figure 10.19 – Selecting LuaLaTeX

6. Compile, and see what we get:

The quick brown fox jumps over the lazy dog. 1234567890

Figure 10.20 – Segoe UI Semilight from Microsoft Windows 10

That font selection was straightforward: load one package and use one command. Let's now do it with multiple fonts in a document.

Selecting multiple font families

Windows already comes with a number of fonts. We can pick a few in **Settings** | **Fonts** in the Windows **Start** menu, or by looking in the `C:\Windows\Fonts` folder. This time, we will choose the following:

- Cambria as the main serif font
- Segoe UI as the sans-serif font
- Lucida Console as the typewriter font
- Cambria Math as the math font

All of these are standard Windows fonts.

Let's assemble a document that displays all four fonts:

1. Start a new document and enter our `\pangram` macro again for easy testing:

```
\documentclass{article}
\newcommand{\pangram}[1]{{#1 The quick brown fox
jumps over the lazy dog. 1234567890\par}}
```

2. Load the `fontspec` package and the `unicode-math` package. The latter lets us select the math font:

```
\usepackage{fontspec}
\usepackage{unicode-math}
```

3. Set the fonts as planned using an optional argument that automatically scales the fonts, so their lowercase letter height matches the main font's lowercase letter height, as follows:

```
\setmainfont{Cambria}
\setsansfont{Segoe UI}[Scale=MatchLowercase]
\setmonofont{Lucida Console}[Scale=MatchLowercase]
\setmathfont{Cambria Math}[Scale=MatchLowercase]
```

4. Then, write a test document body again to view the available fonts:

```
\begin{document}
\large
\pangram{\rmfamily}
\pangram{\sffamily}
\pangram{\ttfamily}
\[
  \int_a^b \! f(x) \, dx = \lim_{\Delta x \rightarrow 0}
  \sum_{i=1}^{n} f(x_i) \,\Delta x_i
\]
\end{document}
```

5. Compile with **LuaTeX** or **XeLaTeX**, and take a look:

The quick brown fox jumps over the lazy dog. 1234567890
The quick brown fox jumps over the lazy dog. 1234567890
`The quick brown fox jumps over the lazy dog. 1234567890`

$$\int_a^b f(x)\,dx = \lim_{\Delta x \rightarrow 0} \sum_{i=1}^{n} f(x_i)\,\Delta x_i$$

Figure 10.21 – Various Microsoft Windows fonts

Cambria became the main text font, and whenever we switch to sans-serif, we get Segoe UI, and when we write code listings in typewriter font, Lucida Console appears. Also, math formulas are now printed in Cambria Math rather than the default Computer Modern font. This easy font selection is actually quite an evolution in LaTeX, and it's worth considering LuaLaTeX or XeLaTeX just because of the extended font support. Both support OpenType and TrueType fonts, but neither works with **pdfLaTeX** yet.

XeLaTeX was developed with direct system-font access in mind, something pdfLaTeX doesn't offer. LuaLaTeX was introduced as a LaTeX extension that added the Lua scripting language, and it has since gained better font support. Without dealing with their advanced features, we can simply choose one of them for fonts when we don't have a package for pdfLaTeX.

Nowadays, XeLaTeX is less maintained, and it's recommended to use LuaLaTeX.

Selecting a font locally

As promised earlier, we'll also work with macOS. This time, let's choose the **Zapfino** font, which is preinstalled in macOS. It's a beautiful calligraphic font we'll apply only to a specific piece of text, rather than changing the font for the entire document. Follow these steps:

1. Start with a small document, just like in the previous sections. As before, make sure you load the `fontspec` package:

```
\usepackage{fontspec}
```

2. Define a new font family, with a name of your choice, and set it to `Zapfino`:

```
\newfontfamily{\calligraphicfont}{Zapfino}
```

3. Define a custom formatting macro, as we did in *Chapter 2, Formatting Text and Creating Macros*:

```
\newcommand{\calligraphic}[1]{{\calligraphicfont #1}}
```

4. Now, you can use this command anywhere in your document text to format it calligraphically:

```
\calligraphic{A graceful Zapfino typography example}
```

5. Check the output of this macro:

Figure 10.22 – Selecting a calligraphic font

Here, ascenders and descenders are intentionally exaggerated to look more elegant. This change affects only the selected text.

Summary

You now know how to choose and combine different text and math fonts, so your documents no longer have to rely on LaTeX's default look.

We explored complete font sets, individual fonts, and several good font packages. For some advanced font techniques with ready-to-use examples, you can take a look at *Chapter 3, Adjusting Fonts*, in the *LaTeX Cookbook*, and visit the book's website with online compilable code at `https://latex-cookbook.net/chapter-03`.

Now, let's go from fonts back to LaTeX in general, and we will learn how to develop and manage larger documents in the next chapter.

11

Developing Large Documents

The first chapter of this book mentioned that LaTeX can handle large documents with ease. Once you start creating larger documents, you'll see that LaTeX continues to work reliably no matter how big the document becomes. For the computer, the source code structure doesn't matter, but for you, as the author, it's important to keep things organized. Long documents can span hundreds of pages with thousands of lines, so a manageable structure is essential.

By the end of this chapter, you'll be able to set up a large document project with multiple files, a title page, and separately numbered front matter and back matter.

In this chapter, we'll look at the following topics:

- Splitting the input
- Creating front and back matter
- Designing a title page
- Working with templates
- Organizing files and folders

That's a big step toward writing a thesis, a book, or a comprehensive report.

Let's begin by building a document from several files.

Splitting the input

To keep a large document manageable, we can split it into several smaller files. This lets us work on chapters separately while maintaining a clean overall structure.

We'll separate the preamble from the body text and place each chapter in its own file. As an example project, we'll build a multi-chapter document on equations and systems of equations, similar in style to a thesis or a book. We can reuse material from the example in *Chapter 9, Writing Math Formulas*, where we worked with theorems about equations.

We'll create several files, step by step:

1. Start a new file, and collect all the package loads and their options, as we did in our preambles in the previous chapters. This can be a lot of packages that we have already learned about, like this:

   ```
   \usepackage[english]{babel}
   \usepackage[T1]{fontenc}
   \usepackage{lmodern}
   \usepackage{microtype}
   \usepackage{natbib}
   \usepackage{tocbibind}
   \usepackage{amsmath}
   \usepackage{amsthm}
   \newtheorem{thm}{Theorem}[chapter]
   \newtheorem{lem}[thm]{Lemma}
   \theoremstyle{definition}
   \newtheorem{dfn}[thm]{Definition}
   ```

2. Save this file as `preamble.tex`.

3. Start another new document and copy the contents of the *Equations* chapter from the theorem example of the *Writing theorems and definitions* section of *Chapter 9, Writing Math Formulas*:

   ```
   \chapter{Equations}
   \section{Quadratic equations}
   \begin{dfn}
     A quadratic equation is an equation of the form
     \begin{equation}
       \label{quad}
       ax^2 + bx + c = 0
     \end{equation}
     where \( a, b \) and \( c \) are constants
     and \( a \neq 0 \).
     \end{dfn}
   ```

4. Save this document as `chapter1.tex`.
5. Create another document for the next chapter, and write the chapter heading and some more text, including a few sections. Save it as `chapter2.tex`:

```
\chapter{Equation Systems}
\section{Linear Systems}
...
\section{Non-linear Systems}
...
```

6. Now, create the top-level document, `equations.tex`. This one starts with the `\documentclass` command and lists the preamble and the chapters for inclusion:

```
\documentclass{book}
\input{preamble}
\begin{document}
\tableofcontents
\include{chapter1}
\include{chapter2}
\end{document}
```

7. Compile the document twice. Remember that this action is necessary to finally get the table of contents. Check the contents to see whether everything is in its correct place:

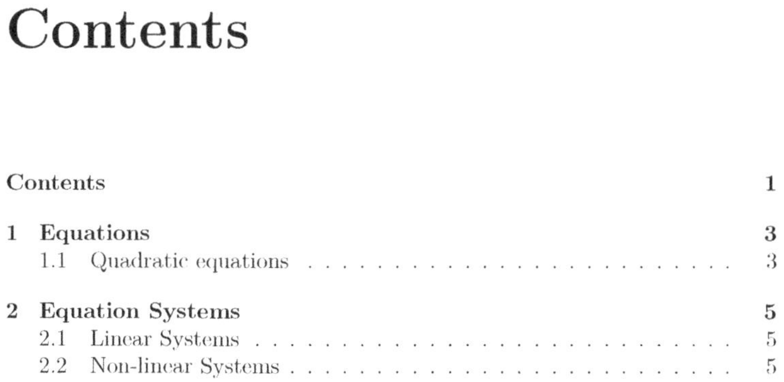

Contents

Contents 1

1 Equations 3
1.1 Quadratic equations 3

2 Equation Systems 5
2.1 Linear Systems 5
2.2 Non-linear Systems 5

Figure 11.1 – Table of contents

We constructed a top-level document named `equations.tex`. You can choose a meaningful name, since the filename determines the resulting PDF's name.

This `.tex` file serves as the framework for our project. It's a regular LaTeX document, but we reduced it as much as possible and used two commands to import external `.tex` files:

- The `\input` command inserts another file, just as if we had typed it directly
- The `\include` command also reads an external file, but automatically inserts `\clearpage` before it

Let's compare these commands in detail.

Including small pieces of code

The simplest way to read another file in your document is with this command:

```
\input{filename}
```

When LaTeX encounters this command, it reads the contents of the file named `filename`, exactly as if it had been typed directly there. The LaTeX compiler processes all commands in this file.

If no filename extension is given, LaTeX assumes the `.tex` extension and inserts `filename.tex`. You can also specify a path, either relative or absolute. Since a backslash starts a command, use forward slashes (/) instead of backslashes (\) in path names. For example, use `\input{chapters/ch01.tex}` instead of `\input{chapters\ch01.tex}`.

Using relative paths makes it easier to move or copy a project. Avoid special characters and spaces in filenames and paths; stick to letters, numbers, -, and _.

Use `\input` to put your preamble in a separate file. It keeps your main document tidy, and you can easily copy and adapt a separate preamble for your other projects.

However, splitting a document with `\input` is not always enough for managing large projects. You might comment out individual `\input` lines to compile only parts of a large document, especially when complete compilation takes time. But doing so would disrupt the page numbering, section numbering, and cross-referencing. The `\include` command is designed to handle these issues, so let's take a closer look at it next.

Including bigger parts of a document

For bringing in larger sections of a document, such as whole chapters, the following command is more suitable:

```
\include{filename}
```

The argument is handled similarly to `\input`, but there are key differences:

- `\include` implicitly starts a new page. Simplified, `\include{filename}` behaves like this:

  ```
  \clearpage
  \input{filename}
  \clearpage
  ```

  This makes `\include` useful for page ranges such as chapters or sections. One consequence is that you may use `\include` only after `\begin{document}`.

- `\include` cannot be nested. You could still use `\input` in included documents, although adding too many layers can make the structure harder to manage.
- Most importantly, `\include` supports a mechanism for choosing which parts of the document you wish to compile, which is specified using another command, namely, `\includeonly`.

Let's see how `\includeonly` works.

Compiling parts of a document

Such a partial document, intended for `\input` or `\include`, cannot be compiled on its own; it needs a root document that specifies the document class.

However, once you swap out parts of the document using `\include` while compiling your root document, you may choose which sections or chapters are included with this command:

```
\includeonly{file list}
```

The `\includeonly` command must appear in the preamble, that is, before `\begin{document}`.

The argument can be a single filename or a comma-separated list of filenames. If a file named `name.tex` is not specified in this argument, `\include{name}` will not insert the file; it will act just like a `\clearpage` command. This allows the exclusion of chunks or whole chapters from compiling. If you work on a large document, this speeds up compilation when you include only your current chapter, while keeping the labels and references intact.

As with the `main.tex` file, LaTeX writes an `.aux` file for each included `.tex` file and reads all of them during compilation, since they contain information such as chapter and page numbers. As long as each included file has been compiled at least once, cross-references and the numbering of pages, chapters, sections, and so on will remain correct, even when you temporarily exclude some chapters.

Try it out by adding the following:

```
\includeonly{chapter2}
```

Add it to your preamble in `equations.tex` and compile. The result will be just the second chapter, keeping the correct numbering. Take a look at the output here with Acrobat Reader:

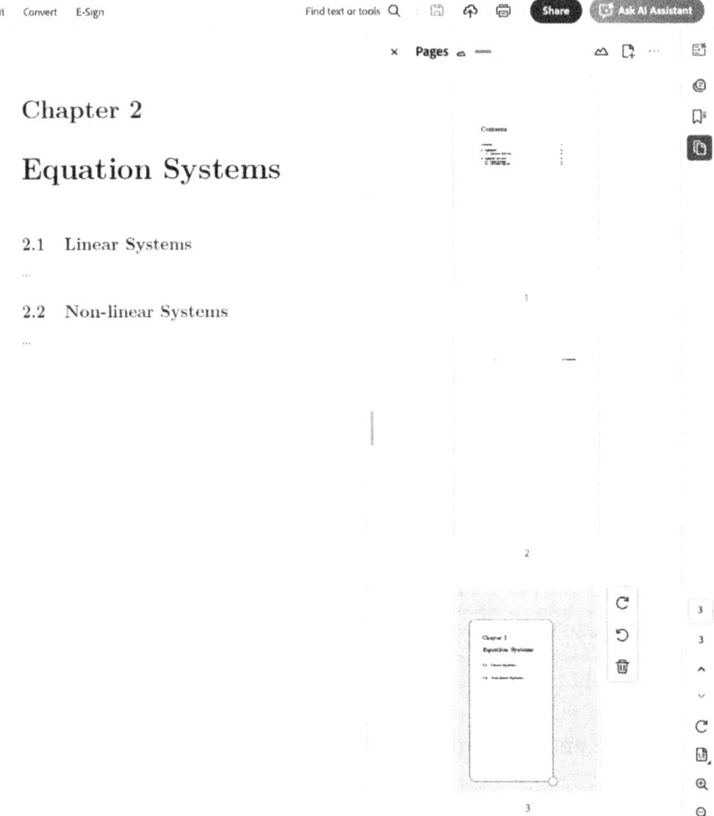

Figure 11.2 – Our document with Chapter 2 only

At the top of *Figure 11.2*, you can see **3/3** pages instead of **5** pages in the first example of the current chapter. On the side, you see three thumbnail pages, and the third is our *Chapter 2*. This shows us that *Chapter 1, Equations,* is not included here, but only *Chapter 2*, as we desired. Also, the page numbers remain the same as in the full document, and the original chapter and section numbers are unchanged.

Compiling time is significantly shorter if you work on a huge document with many chapters and use `\includeonly` to include just a single chapter that you are working on.

Finally, comment out the `\includeonly` command to typeset your complete document when you finish your work.

Of course, you can use `\include` without `\includeonly` just for splitting a large document into files.

Let's now return to the structure of a bigger document.

Creating front and back matter

Books often begin with introductory material such as copyright information, a foreword, acknowledgments, or a dedication. Similar for a thesis. This part, including the title page and the table of contents, is called the **front matter**. A thesis can follow the same structure.

At the end, a book or thesis can include an afterword and supporting material such as a bibliography and an index. This part is called the **back matter**.

The `book` class and other similar classes, such as `scrbook` and `memoir`, support this sectioning directly. They also support different numbering styles for pages and chapters. Let's see how it works.

In our example, the book begins with a dedication. The front matter will include the table of contents, the list of tables and figures, and a dedication. All the pages of the front matter will be numbered with Roman numerals. At the end, we add an appendix with supplementary proofs that we like to keep separate from the main chapters:

1. Create a file called `dedication.tex`:

   ```
   \chapter{Dedication}
   This book is dedicated to one of the greatest
   mathematicians of all time: Carl Friedrich Gauss.
   Without him, this book wouldn't have been possible.
   ```

2. Create a file called `proofs.tex`:

   ```
   \chapter{Proofs}
   ...
   ```

3. Extend the main file, `equations.tex`, by means of the highlighted lines:

   ```
   \documentclass{book}
   \input{preamble}
   \begin{document}
   ```

```
\frontmatter
\include{dedication}
\tableofcontents
\listoftables
\listoffigures
\mainmatter
\include{chapter1}
\include{chapter2}
\backmatter
\include{proofs}
\nocite{*}
\bibliographystyle{plainnat}
\bibliography{example}
\end{document}
```

4. As you can see in the last highlighted line, we reused the `example.bib` file from *Chapter 8, Managing Contents, Indexes, and Bibliography*. Compile, run BibTeX, and compile again. Check out the numbering within the table of contents:

Contents

Figure 11.3 – Table of contents of a complex document

We saw that LaTeX printed the page number of the contents page in Roman numerals. This applies to all front-matter pages. Furthermore, all the chapters in the front and back matter are unnumbered, even though we did not use the starred command, `\chapter*`.

This behavior is controlled by three commands, namely, `\frontmatter`, `\mainmatter`, and `\backmatter`. Each starts a new page and adjusts both the page and chapter numbering as follows:

- `\frontmatter`: Pages are numbered with lowercase Roman numerals. Chapters generate a table of contents entry, but don't get a number.
- `\mainmatter`: Page numbers switch to Arabic numerals. Chapters are numbered and produce table-of-contents entries.
- `\backmatter`: Pages continue with Arabic numerals. Chapters appear again in the table of contents entry but are not numbered.

Like the book class, the `scrbook` and `memoir` classes provide the same commands with very similar behavior.

A large document usually begins with a title page, so let's look at how to create one in LaTeX.

Designing a title page

You can create a good-looking title page quickly using the `\maketitle` command, as we did in *Chapter 2, Formatting Text and Creating Macros*. Most document classes offer this command to generate a reasonable, pre-formatted title page. If you want complete control over the layout, you could use a `titlepage` environment instead. So, let's design a nice title page for our book of equations.

In *Chapter 2*, we already used some formatting commands, such as `\centering`, and font size and shape commands, such as `\Huge` and `\bfseries`, to format a title. We will do it similarly within a `titlepage` environment:

1. Create a file, `title.tex`, with the following content:

   ```
   \begin{titlepage}
   \raggedleft
   {\Large The Author\\[1in]}
   {\large The Big Book of\\}
   {\Huge\scshape Equations\\[.2in]}
   {\large Packed with hundreds of examples and solutions\\}
   \vfill
   {\itshape 2011, Publishing company}
   \end{titlepage}
   ```

2. Add this line right after `\frontmatter`:

```
\include{title}
```

3. Our final book will be in A5 format, and so will the title page. Therefore, let's add that to the preamble:

```
\usepackage[a5paper]{geometry}
```

4. Compile; now we've got a title page:

The Author

The Big Book of

EQUATIONS

Packed with hundreds of examples and solutions

2011, Publishing company

Figure 11.4 – A title page

The `titlepage` environment places its contents on a separate page. Although this title page will be numbered like any other page, the page number won't be printed on that page.

In this environment, we used basic LaTeX font commands to modify the font size and style. By grouping with curly braces, we limited the scope of those commands. In square brackets after line breaks such as `\\[.2in]`, we added a bit of extra vertical space before the following line. The `\vfill` command inserts a flexible vertical space that stretches as far as possible to fill the entire page. This way, we pushed the last line toward the bottom of the page.

Note that this page has the same page dimensions as the other pages in the document. That means, in a double-sided book, that it's a right-hand page. Thus, you may notice the unequal left and right margins, which might look undesirable, especially if your title is in the center. However, the explanation is simple: this title page is intended to be an **inner title**, not the **cover page**. The inner title page is, of course, a right-hand page.

A cover page is different. Such a page should be one-sided, and thus it should have equal left and right margins. A cover page is often produced as a standalone document and printed separately. For an electronic document, you may use the `pdfpages` package. Refer to the *Including whole pages* section in *Chapter 6, Including Images*, or the *Combining PDF files* section in *Chapter 9, Optimizing PDF files*, in the *LaTeX Cookbook*, and to `https://texdoc.org/pkg/pdfpages`.

The `titling` package offers features to create sophisticated title pages. To get some ideas on how title pages may be designed, you could look at *Some Examples of Title Pages*, by Peter Wilson, available from `texdoc titlepages` and at `https://texdoc.org/pkg/titlepages`.

A document framework consisting of files, headings, a title page, and style settings is called a **template**. We will learn how to use templates in the next section.

Working with templates

When creating a document, we specify the document class, choose appropriate packages and options, and establish a structure for the content. Repeating these steps for every new document quickly becomes tedious.

If you plan to produce multiple documents of the same type, you may create a template. This could be a `.tex` file containing the following:

- A declaration of a suitable document class together with meaningful options
- Routinely used packages that are most suitable for your document type
- A predefined layout for the header, footer, and main text
- Self-made macros to streamline your work
- A framework of sectioning commands, where you fill in the headings and the body text
- A framework containing `\include` or `\input` commands, for which you can create the body text chunks later

As your LaTeX skills grow, your templates will evolve and become better and more sophisticated. Many users publish their elaborate templates online. Many universities, institutes, journals, and publishers do the same, offering official templates for documents such as theses, papers, journal articles, and books that match their formatting requirements.

You can find a carefully curated collection of templates, arranged by document types such as theses, reports, letters, and presentations, along with sample outputs, in a template gallery at `https://latextemplates.com`.

You may download a template and start to fill in your text. Alternatively, you could start a document with a predefined template offered by your editor. Let's try that first.

LaTeX editors often provide templates to start with. TeXworks offers some as well. So, we'll test this feature. Let's take one, open it, modify it, and compile it:

1. In the TeXworks main menu, click **File**, then **New from template**. A window will open, allowing you to choose a template:

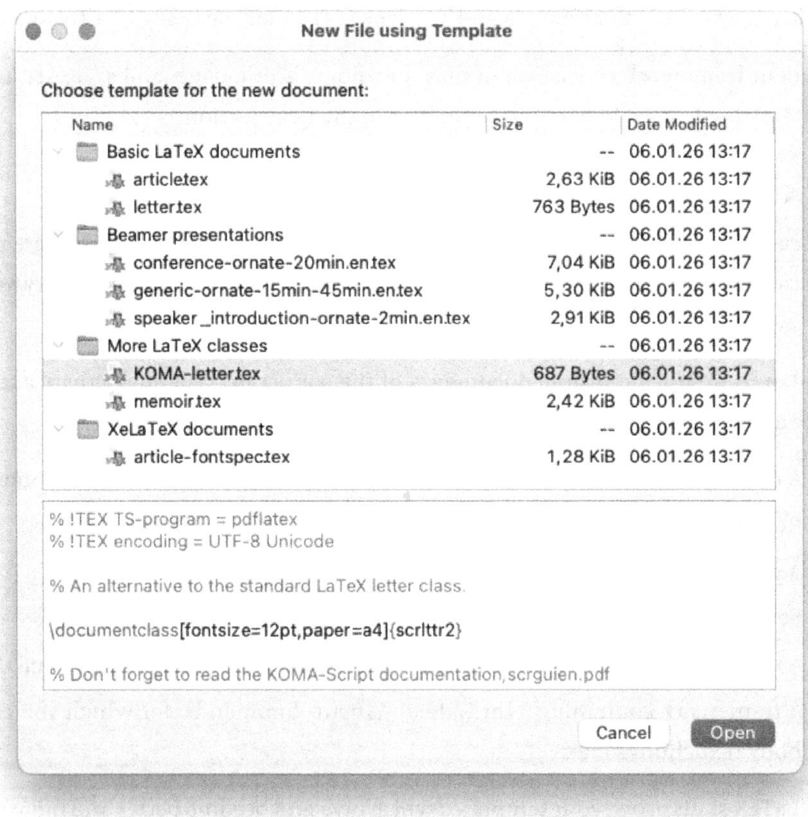

Figure 11.5 – TeXworks template selection

2. In the lower part of the window, you can read the template's source. Here's an example of KOMA-Script (`KOMA-letter.tex`):

```
% !TEX TS-program = pdflatex
% !TEX encoding = UTF-8 Unicode
% An alternative to the standard LaTeX letter class.
\documentclass[fontsize=12pt, paper=a4]{scrlttr2}
% Don't forget to read the KOMA-Script documentation,
% scrguien.pdf
\setkomavar{fromname}{} % your name
\setkomavar{fromaddress}{Address \\ of \\ Sender}
\setkomavar{signature}{} % printed after the \closing
\renewcommand{\raggedsignature}{\raggedright} % make
% the signature ragged right
\setkomavar{subject}{} % subject of the letter
\begin{document}
\begin{letter}{Name and \\ Address \\ of \\ Recipient}
\opening{} % eg. Hello
\closing{} %eg. Regards
\end{letter}
\end{document}
```

3. Click on **Open**. Fill in the gaps and edit the filler text of the example:

```
\documentclass[fontsize=12pt, paper=a4]{scrlttr2}
\setkomavar{fromname}{My name} % your name
\setkomavar{fromaddress}{Street, City}
\setkomavar{signature}{Name} % printed after the \closing
\setkomavar{subject}{Invoice 1/2021} % subject of the letter
\setkomavar{place}{Place}
\setkomavar{date}{January 1, 2021}
\begin{document}
\begin{letter}{Customer Name\\ Street No. X \\ City \\ Zipcode}
  \opening{To whom it may concern} % eg. Hello
  Text follows \ldots
  \bigskip
  \closing{With kind regards} %eg. Regards
\end{letter}
\end{document}
```

4. Compile the document. Have a look at our test letter:

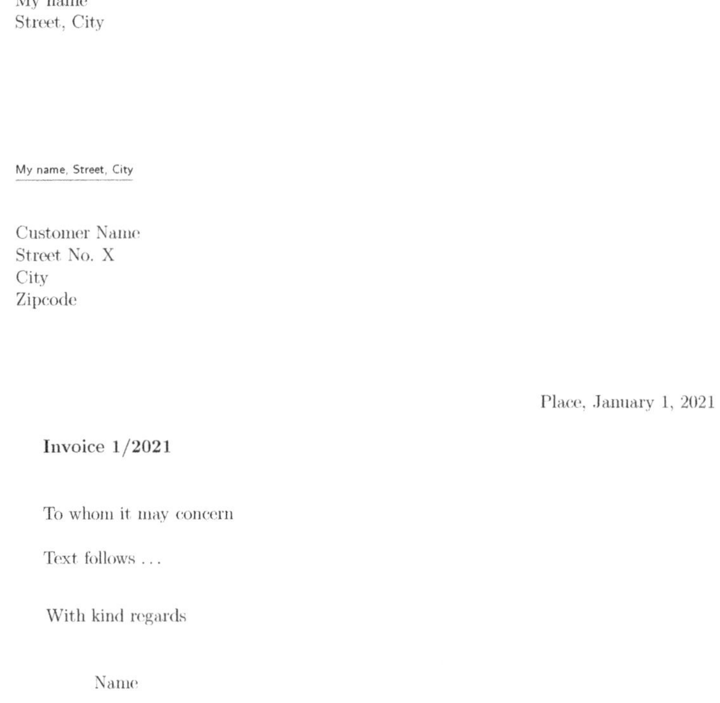
My name
Street, City

My name, Street, City

Customer Name
Street No. X
City
Zipcode

Place, January 1, 2021

Invoice 1/2021

To whom it may concern

Text follows ...

With kind regards

Name

Figure 11.6 – A letter document

That was easy! We just opened the template and modified the placeholder text. By reading the KOMA-Script documentation, we can learn that the `\setkomavar` command is used to specify values for template parameters such as `name`, `address`, and `subject`. We used that to declare `date` and `place` as well.

Once you've written your personal data into this template, you may save that for later use instead of typing our address for each letter.

The KOMA-Script documentation, available via `texdoc koma-script` and at `https://texdoc.org/pkg/koma-script`, describes the features of the `scrlttr2` class well. Using this, you can create a professional-looking business letter template.

Imagine putting a job application letter created with the LaTeX layout and fonts together with the `microtype` package next to an application letter produced with some other word processing software. Which one will create a better impression?

While looking for LaTeX templates, code, and tips on the internet, you will encounter a lot of information and code. This code might be outdated, and the information might be obsolete.

When you develop your own template, you probably want to use the best packages, options, and modern solutions available today. How can you be sure?

Both questions can be answered by studying the `l2tabu` document. This name is the common shortcut for *An essential guide to LaTeX2e usage*. This document highlights obsolete commands and packages and demonstrates the most common and severe mistakes LaTeX users make. As LaTeX has developed over many years, some packages and techniques are still available and described in online resources, but they may no longer be recommended. Read this guide. It will help you evaluate the templates and code you find online and ensure that you produce optimal code yourself.

Just type `texdoc l2tabuen` (*en* stands for *English*) at the command prompt or visit `https://texdoc.org/pkg/l2tabuen`.

To test a template, you may use the `blindtext` package and its commands, `\blindtext` and `\Blinddocument`. The `\blindtext` command generates a paragraph of dummy text, and the `\Blinddocument` command generates dummy content for a big document, including sections and lists. That will demonstrate the template's output quality. When using this package, we should load the `babel` package with a language option, for example, just with a basic minimal article template:

```
\documentclass{article}
\usepackage[english]{babel}
\usepackage{blindtext}
\begin{document}
\begin{abstract}
\blindtext
\end{abstract}
\Blinddocument
\end{document}
```

That gives us a document starting like this:

Abstract

Hello, here is some text without a meaning. This text should show what a printed text will look like at this place. If you read this text, you will get no information. Really? Is there no information? Is there a difference between this text and some nonsense like "Huardest gefburn"? Kjift – not at all! A blind text like this gives you information about the selected font, how the letters are written and an impression of the look. This text should contain all letters of the alphabet and it should be written in of the original language. There is no need for special content, but the length of words should match the language.

1 Heading on Level 1 (section)

Hello, here is some text without a meaning. This text should show what a printed text will look like at this place. If you read this text, you will get no information. Really? Is there no information? Is there a difference between this text and some nonsense like "Huardest gefburn"? Kjift – not at all! A blind text like this gives you information about the selected font, how the letters are written and an impression of the look. This text should contain all letters of the alphabet and it should be written in of the original language. There is no need for special content, but the length of words should match the language.

This is the second paragraph. Hello, here is some text without a meaning. This text should show what a printed text will look like at this place. If you read this text, you will get no information. Really? Is there no information? Is there a difference between this text and some nonsense like "Huardest gefburn"? Kjift – not at all! A blind text like this gives you information about the selected font, how the letters are written and an impression of the look. This text should contain all letters of the alphabet and it should be written in of the original language. There is no need for special content, but the length of words should match the language.

And after the second paragraph follows the third paragraph. Hello, here is some text without a meaning. This text should show what a printed text will look like at this place. If you read this text, you will get no information. Really? Is there no information? Is there a difference between this text and some nonsense like "Huardest gefburn"? Kjift – not at all! A blind text like this gives you information about the selected font, how the letters are written and an impression of the look. This text should contain all letters of the alphabet and it should be written in of the original language. There is no need for special content, but the length of words should match the language.

After this fourth paragraph, we start a new paragraph sequence. Hello, here is some text without a meaning. This text should show what a printed text will look like at this place. If you read this text, you will get no information. Really? Is there no information? Is there a difference between this text and some nonsense like "Huardest gefburn"? Kjift – not at all! A blind text like this gives you information about the selected font, how the letters are written and an impression of the look. This text should contain all letters of the alphabet

Figure 11.7 – An article with dummy text

If you use the TeXworks editor, as we did in the previous example, you can choose from some ready-to-use templates or download a template from `https://latextemplates.com`. However, if you work online with `https://overleaf.com`, you have even more choices. Basically, you can click on **New Project** and choose one of several basic templates:

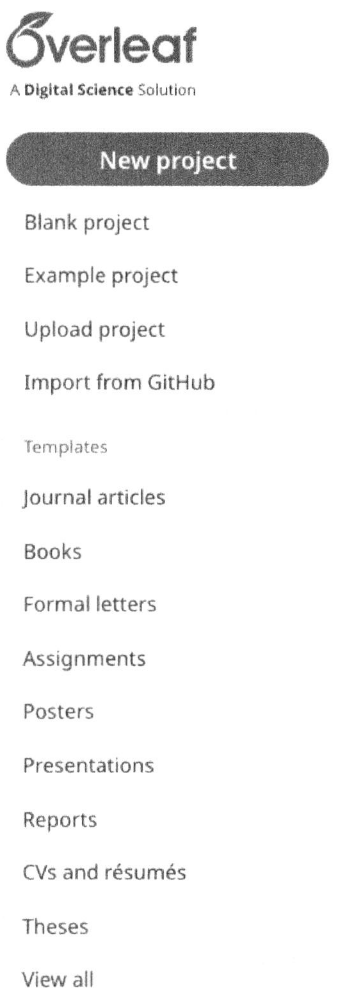

Figure 11.8 – Opening a template in Overleaf

If you click on **View All**, you can browse a comprehensive catalog:

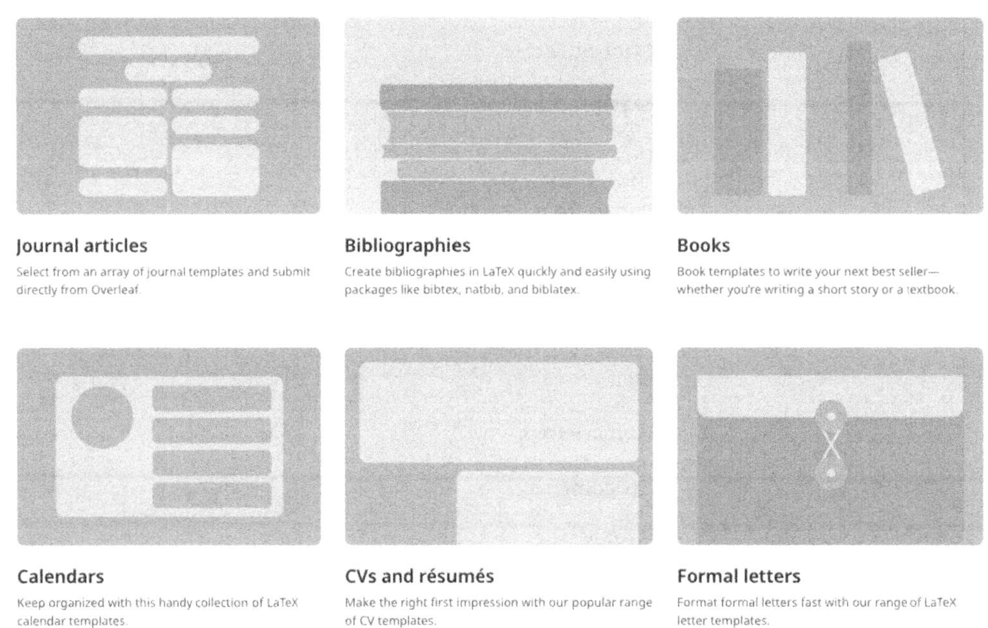

Figure 11.9 – The Overleaf template catalog

The Overleaf template collection contains several thousand ready-to-use templates for filling in your text. Institutions and users contribute a lot of them. Quality may vary, but you can see screenshots of the contents or titles while browsing and try out those templates yourself.

You can enter keywords such as the document type, university or college name, features, or package names in the **Search** bar.

You can click on **Open as Template** and get a compilable document with some filler text. That lets you try out about 10 templates in around 10 minutes until you find the perfect one.

Some templates are fairly complex, with content split across multiple files and arranged in folders. Let's look at how we can set up such a structure ourselves next.

Organizing files and folders

Earlier in this chapter, we learned how to split a document into multiple files. As a project grows (think of a thesis or a book), you will quickly accumulate many of them: chapter `.tex` files, appendices, front and back matter, bibliography files, figures in various formats, and perhaps files with custom macros as well.

Keeping everything in a single folder quickly becomes hard to manage. A better approach is to group related files into folders and subfolders. This makes the project easier to navigate and maintain as it grows, especially because LaTeX also creates auxiliary files in your project folder.

Here's a simple example structure with placeholder names to show how you might set things up:

```
project/
    preamble.tex
    macros.tex
    main.tex
    university-style.sty
    frontmatter/
        titlepage.tex
        abstract.tex
        preface.tex
        acknowledgements.tex
    chapters/
        chapter01-intro.tex
        chapter02-methods.tex
        ...
    figures/
        logo.png
        chapter01/
            plot.png
            diagram.png
        chapter02/
        ...
    tables/
        ...
```

```
    appendices/
        appendixA.tex
        appendixB.tex
    backmatter/
        bibliography.tex
        indexes.tex
        declaration.tex
```

Such a project structure gives you better control over all parts of your content. Good naming conventions, such as numeric prefixes and descriptive filenames, help too.

Your `main.tex` file may then look something like this:

```
\documentclass[12pt,a4paper]{report}
\usepackage{university-style}
\input{preamble}
\input{macros}
\usepackage{graphicx}
\graphicspath{
  {figures/}
  {figures/chapter01/}
  {figures/chapter02/}
}
\begin{document}
\input{frontmatter/titlepage}
\input{frontmatter/abstract}
\input{frontmatter/preface}
\input{frontmatter/acknowledgements}
\include{chapters/chapter01-intro}
\include{chapters/chapter02-methods}
\appendix
\include{appendices/appendixA}
\include{appendices/appendixB}
\input{backmatter/bibliography}
\input{backmatter/indexes}
\input{backmatter/declaration}
\end{document}
```

The `\graphicspath` command specifies where LaTeX looks for image files. By defining these paths once in the preamble, you can include graphics without repeating long directory names.

Now, you can concentrate on most of your work in the chapter files.

From time to time, make a backup of the entire project folder, such as a `project.zip` file, and store it elsewhere in case your computer is damaged. For larger projects, it's a good idea to use version control, such as **Git**, to keep track of changes over time. If you're new to it, the official Git documentation at `https://git-scm.com` is a clear, beginner-friendly starting point.

Summary

The techniques introduced in this chapter help you create and maintain bigger projects. Many people learn LaTeX mainly because they plan to write longer texts, such as a thesis or a book. However, splitting documents and using templates are helpful for short pieces of writing as well, such as letters; think of the header, footer, and address fields.

In this chapter, we created and managed large documents consisting of several files, including front and back matter and a separate title page, and organized them in a good folder structure.

Now that we can develop and handle large documents, we will see how to improve them further in the next chapter.

12

Using Hyperlinks and Designing Headings

In the previous chapter, you learned how to structure and manage large documents. With all that you learned up to this point, you can produce well-structured documents with fine typographical quality, suitable for books, journal articles, or a university thesis.

If you plan to publish your work online, your PDF document would benefit from features such as hyperlinks and bookmarks that support navigation.

In this chapter, we will work with tools for such navigation enhancements and for customizing headings and colors in the following sections:

- Working with hyperlinks
- Creating bookmarks
- Designing headings

Each topic is handled with dedicated LaTeX packages, which we will now explore.

Working with hyperlinks

There's a sophisticated package called `hyperref` that automatically adds hyperlinking features to nearly every structural element of a LaTeX document. Let's see how it works.

Adding hyperlinks

Load the `hyperref` package and inspect the result:

1. Open the `preamble.tex` file, which we used in the previous chapter. At the end, add this line:

```
\usepackage{hyperref}
```

2. Save this document under the same name.
3. Open our *Book of Equations* from the previous chapter; we called it `equations.tex`.
4. Compile the document twice without making any changes. Let's see how the document now appears; here, we can see red boxes indicating hyperlinks:

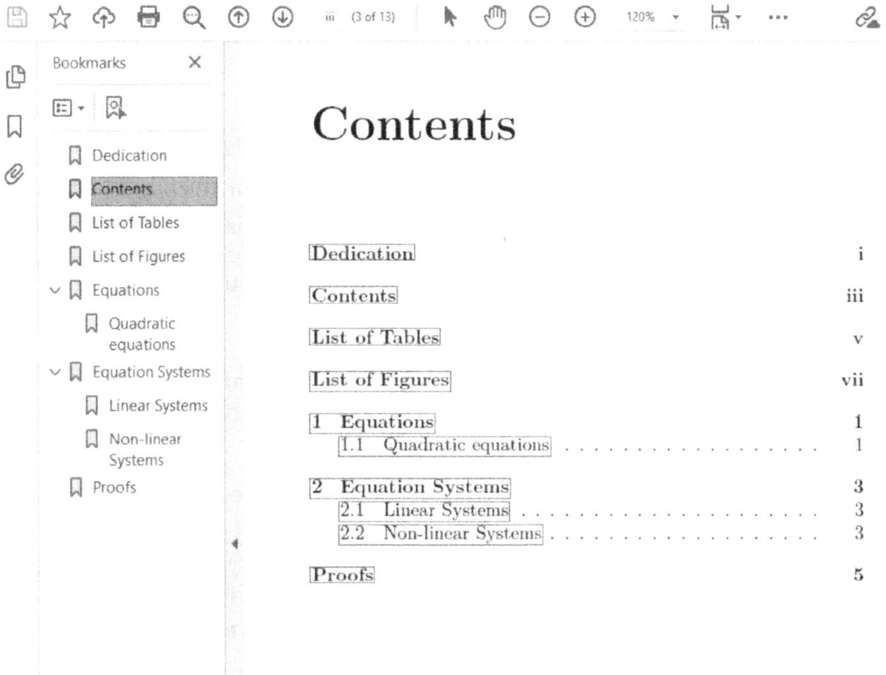

Figure 12.1 – A table of contents with hyperlinks and bookmarks

Cross-references, such as references to equation numbers, also have red boxes:

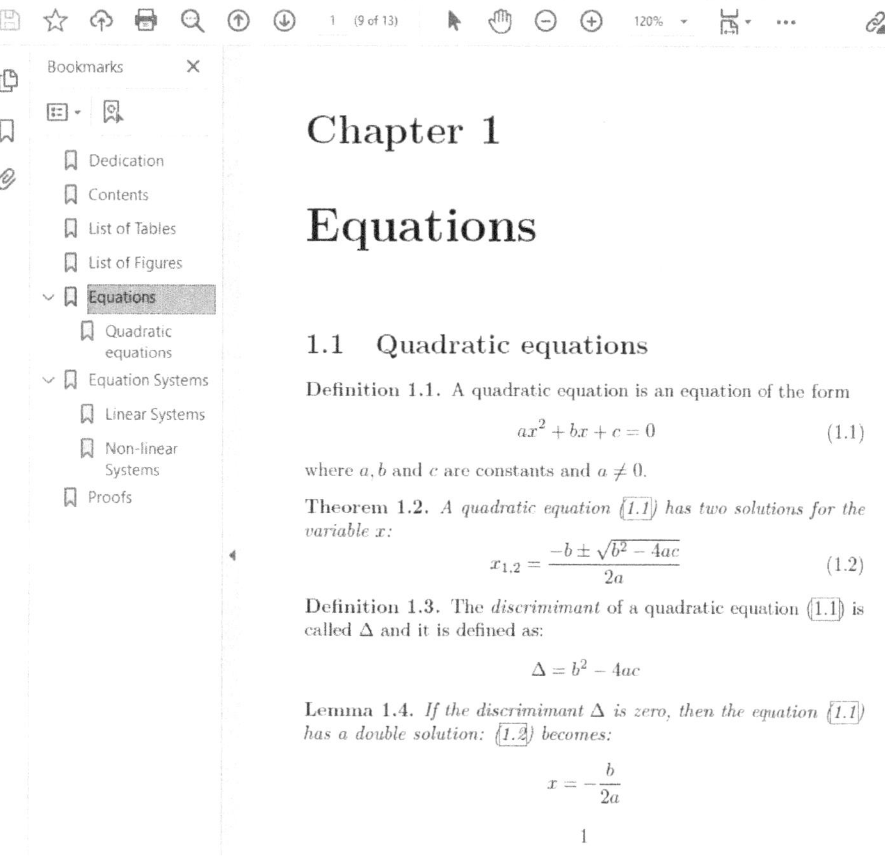

Figure 12.2 – References to equations with hyperlinks

By just loading the `hyperref` package, our document has significantly changed:

- We got a **Bookmarks** bar that allows us to navigate the document easily
- The document contains hyperlinks that are highlighted by red borders
- Each entry in the table of contents has become a hyperlink to the beginning of the corresponding chapter
- All cross-references have become hyperlinks

This is an excellent improvement for electronic versions of our documents.

The red boxes will not appear on paper when you print the document; they appear only on screen for electronic navigation. The same applies to the bookmarks.

If you don't like the default red border for hyperlinks, you can easily change it by editing the `hyperref` options. Let's try this next.

Customizing hyperlinks

Now, we'll pass options to the `hyperref` package that modify the way links are emphasized:

1. Open the `preamble.tex` file again. This time, specify the options for `hyperref`:

```
\usepackage[colorlinks=true,linkcolor=red]{hyperref}
```

2. Save this document, go to the main `equations.tex` document, and compile it twice. The table of contents has changed:

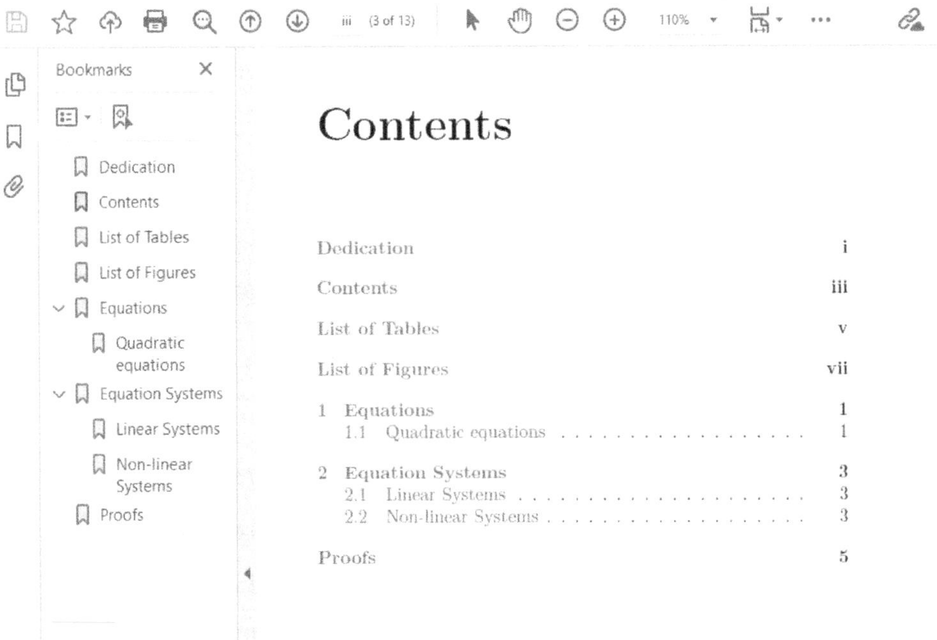

Figure 12.3 – Table of contents with colored hyperlinks

Instead of frames, we now use red for emphasized hyperlinks. Unlike the boxes, we can see the color in a printed document.

`hyperref` provides ways to configure these options. The first one we used is the following:

```
\usepackage[key=value list]{hyperref}
```

Alternatively, we could write \usepackage{hyperref} and set the options afterward:

```
\hypersetup{key=value list}
```

Our example would do the same with the following:

```
\hypersetup{colorlinks=true,linkcolor=red}
```

We can also combine these methods.

Let's look at a selection of handy options. For the following options, choose true or false. If you don't specify a value, hyperref selects the default one, which is shown in parentheses here:

- draft: Turns all hypertext options off (false)
- final: Turns all hypertext options on (true)
- debug: Prints extra diagnostic messages into the log file (false)
- backref: Adds backlinks to the bibliography, that is, links from bibliography items back to citations in the text (false)
- hyperindex: Adds links to page numbers in the index (true)
- hyperfootnotes: Converts footnote markers into hyperlinks (true)
- hyperfigures: Adds hyperlinks to figures (false)
- linktocpage: Adds links to page numbers instead of to the text in the **table of contents** (**TOC**), **list of figures** (**LOF**), and **list of tables** (**LOT**) (false)
- frenchlinks: Uses small caps for links instead of color (false)
- bookmarks: Writes bookmarks for the PDF reader navigation (true)
- bookmarksopen: Shows all bookmarks in an expanded view when the PDF is opened (false)
- bookmarksnumbered: Includes the section number in bookmarks (false)
- colorlinks: Writes links and anchors in color, depending on the type of link, such as page references, URLs, file references, and citations, instead of printing a border around links (false)

When you enable the `colorlinks` option, you can choose the color for each link type from the following list. Again, the default value is in parentheses:

- `linkcolor`: Color of general links (`red`)
- `citecolor`: Color for citations of bibliography items (`green`)
- `urlcolor`: Color for website addresses, that is, URLs (`magenta`)
- `filecolor`: Color for links to files (`cyan`)

There are many more options for customizing link borders, the PDF page size, anchors, bookmark appearance, and the PDF page display style. The `hyperref` documentation lists them all. Just type `texdoc hyperref` at the command line or visit `https://texdoc.org/pkg/hyperref`.

Suppose you want to turn off all link highlighting, such as for printing on paper; give the `hidelinks` option without a value. Then, links will appear as regular text without borders or colors.

Some text options allow us to specify the metadata of PDF files, such as the author's name, title, and keywords. You can see this information by inspecting the document properties in the PDF reader. This is even more beneficial, as internet search engines can find and classify your PDF document based on this metadata. If you publish online, this improves the chances of readers finding your publication.

That's why we will now add PDF metadata to our *Book of Equations* from *Chapter 11, Developing Large Documents*. In addition to choosing sensible keywords, we will set the title and the author's name. During development, why not choose the great mathematician to whom we dedicated our book? Let's do it:

1. Open the `preamble.tex` file and add the following lines:

   ```
   \hypersetup{pdfauthor={Carl Friedrich Gauss},
     pdftitle={The Big Book of Equations},
     pdfsubject={Solving Equations and Equation Systems},
     pdfkeywords={equations,mathematics}}
   ```

2. Save that file. Go to the main `equations.tex` document and click **Typeset** to compile.
3. Let's inspect the **Document Properties** section. If you are using Acrobat Reader, click on the **File** menu and then on **Properties**:

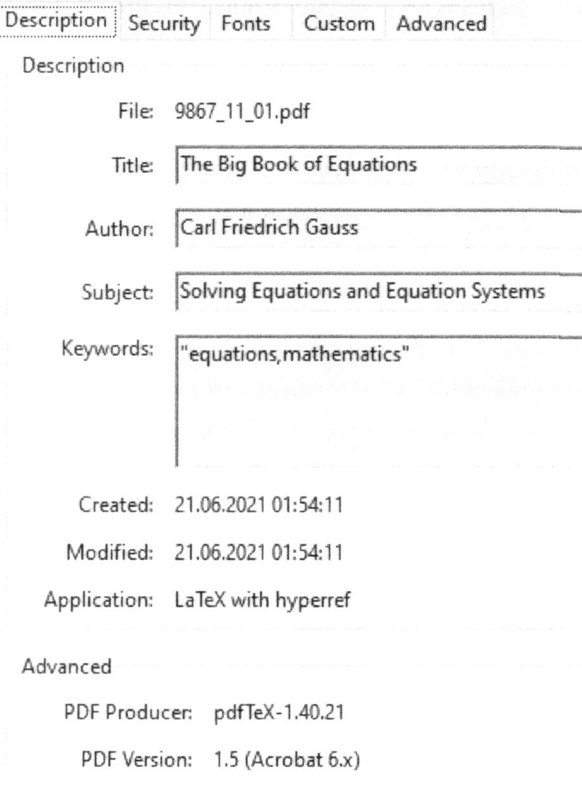

Figure 12.4 – PDF metadata in Document Properties

That was easy. We provided all the document properties using `hyperref` options; we just had to enclose each entry in curly braces.

The most used metadata options are the following:

- `pdftitle`
- `pdfauthor`
- `pdfsubject`
- `pdfcreator`
- `pdfproducer`
- `pdfkeywords`

PDF metadata stores basic information about a document for readers and search engines. The most common metadata fields describe the document itself: `pdftitle`, `pdfauthor`, and `pdfsubject` define the title, author, and overall topic, respectively. The `pdfcreator` and `pdfproducer` fields document the software used to generate and process the PDF, which is usually set automatically but can be customized. Carefully choosing metadata is worthwhile, especially for PDFs shared online. Use a clear, descriptive title instead of a filename, and pick a small, focused set of meaningful keywords. Think about what readers might search for, and avoid vague terms. Good metadata improves searchability and gives your PDF a more professional finish.

Because the `hyperref` package redefines many other packages' commands to add hyperlink functionality, we must load it after those packages. A good rule of thumb is to load the `hyperref` package last in your preamble. A few packages are exceptions to that rule, namely, `algorithm`, `amsrefs`, `bookmark`, `chappg`, `cleveref`, `glossaries`, `hypernat`, `linguex`, `sidecap`, and `tabularx`. For more information, see `https://latexguide.org/hyperref`.

There are further ways to add hyperlinks and bookmarks, which we will see now.

Creating hyperlinks manually

Since the `hyperref` package automatically creates links for nearly all kinds of cross-references, you seldom need to create hyperlinks yourself. Still, the package provides user commands for doing so when needed:

- `\href{URL}{text}` turns text into a hyperlink that points to the URL (the website address)
- `\url{URL}` prints the URL and links it
- `\nolinkurl{URL}` prints the URL without linking it
- `\hyperref{label}{text}` converts the text into a hyperlink that links to the location where the `label` is set (i.e., the same place `\ref{label}` would point to)
- `\hypertarget{name}{text}` creates a target name for potential hyperlinks with `text` as the anchor
- `\hyperlink{name}{text}` turns `text` into a hyperlink that points to the target name

An **anchor** is an internal jump target in the document. Sectioning commands such as `\chapter` or `\section` automatically create anchors, which hyperlinks and TOC entries can point to.

Sometimes, however, you add an entry to the TOC manually using the `\addcontentsline` command, which creates a hyperlinked TOC entry. Still, there hasn't been a corresponding sectioning command setting the anchor. The TOC entry points to the most recently created anchor, which may be the previous chapter or section. As a result, clicking the TOC entry jumps to the wrong page.

The \phantomsection command comes to the rescue; it just sets an anchor like \hypertarget{}{} would do. It's often used this way for creating a TOC entry for the bibliography while linking to the correct page, as follows:

```
\cleardoublepage
\phantomsection
\addcontentsline{toc}{chapter}{Bibliography}
\bibliography{name}
```

In effect, you can think of \phantomsection as inserting an invisible \section anchor. The following \addcontentsline command then refers to that anchor.

Creating bookmarks

Your **Bookmarks** panel might already be full of chapters and section entries created by hyperref. But what if you want to add bookmarks by yourself? You can do that as follows.

\pdfbookmark[level]{text}{name} creates a bookmark with text at the optionally specified level. The default level is 0. Treat the name just like with the \label command; it should be unique because it identifies the internal anchor.

You can also insert bookmarks relative to the current level:

- \currentpdfbookmark{text}{name} places a bookmark at the current level
- \belowpdfbookmark{text}{name} creates a bookmark one level deeper
- \subpdfbookmark{text}{name} steps down one level and creates a bookmark at that deeper level

The bookmark package offers additional customization options, including font style and color. You can read about it by running texdoc bookmark at the command line or visiting https://texdoc.org/pkg/bookmark.

Using math formulas and special symbols in bookmarks

Due to PDF limitations, we cannot use math and special symbols within PDF bookmarks. This becomes an issue, for instance, when a sectioning title contains math symbols or font commands, as hyperref would try to place them into the bookmark. To avoid this, you can provide two versions of the title: one for TeX and one for the PDF bookmark. Use this command:

```
\texorpdfstring{string with TeX code}{pdf text string}
```

It returns the argument depending on the context to avoid such problems. It can be used like this:

```
\section{The equation
  \texorpdfstring{$y=x^2$}{y=x\texttwosuperior}}
```

That may come in handy.

If you load `hyperref` with the `unicode` option, you could use Unicode text characters in bookmarks, such as here:

```
\section{\texorpdfstring{$\gamma$}{\textgamma} radiation}
```

Let's quickly see how these commands work in a small sample document. Here it goes:

```
\documentclass{article}
\usepackage{bm}
\usepackage[colorlinks=true,psdextra,unicode]{hyperref}
\begin{document}
\pdfbookmark[1]{\contentsname}{toc}
\tableofcontents
\pdfbookmark[1]{Abstract}{abstract}
\begin{abstract}
\centering
Sample sections follow.
\end{abstract}
\section{The equation
  \texorpdfstring{$y=x^2$}{y=x\texttwosuperior}}
\section{\texorpdfstring{$\gamma$}{\textgamma} radiation}
\section[\texorpdfstring{Let $\int\sim\sum$ for
  $n\rightarrow\infty$}
  {Let \int\sim\sum\ for n\rightarrow\infty}]
  {Let $\bm{\int\sim\sum}$ for $\bm{n\rightarrow\infty}$}
\end{document}
```

As highlighted in the preceding code, the `\section` command does these three things:

- It prints the section heading. This time, we used the `\bm` command from the `bm` package to produce bold math. Compare it with the other headings.
- It puts the section name into the TOC.

- It creates a bookmark with Unicode text symbols in place of math symbols. We loaded `hyperref` with the `unicode` and `psdextra` options, which allow math symbols in bookmarks.

We get this output with bookmarks:

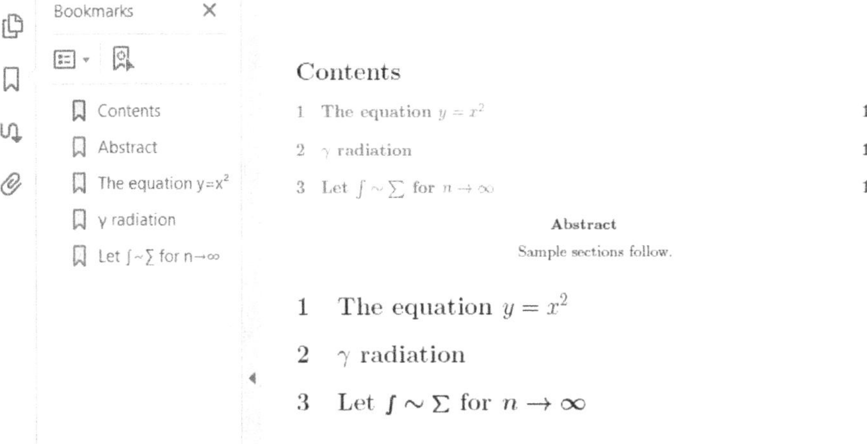

Figure 12.5 – Math formulas in bookmarks

In the first argument to the `\texorpdfstring` command, we used $...$ to enter the math mode. However, in the second argument to `\texorpdfstring`, we omitted $...$ intentionally since that will be Unicode text, not math font glyphs.

While math formulas in headings and bookmarks may not be a good idea anyway, we can do it if we really need to.

In the next section, we will turn to the design and formatting of headings.

Designing headings

In *Chapter 2, Formatting Text and Creating Macros*, we briefly touched on the challenge of customizing headings. There has to be a consistent way to modify the font, spacing, and numbering of headings throughout the document. Fortunately, there's a handy package for that: `titlesec`. We'll use it now to design chapter and section headings.

Let's return to the example that we used earlier in this chapter. We aim to create headings with the following characteristics:

- Centered titles
- A smaller font size
- Reduced space above and below
- A sans-serif font, which is a good choice for bold headings

Let's start:

1. Open the `preamble.tex` file, which we used earlier in this chapter. Insert this line to load the `titlesec` package:

   ```
   \usepackage{titlesec}
   ```

2. Add this command to specify the layout and font of the chapter headings:

   ```
   \titleformat{\chapter}[display]
     {\normalfont\sffamily\Large\bfseries\centering}
     {\chaptertitlename\ \thechapter}{0pt}{\Huge}
   ```

3. Now, define the section heading by calling the `\titleformat` command again:

   ```
   \titleformat{\section}
     {\normalfont\sffamily\large\bfseries\centering}
     {\thesection}{1em}{}
   ```

4. Add this line to adjust the chapter heading's spacing:

   ```
   \titlespacing*{\chapter}{0pt}{30pt}{20pt}
   ```

5. Save `preamble.tex` and compile the main `equation.tex` document. Let's see how the headings have changed:

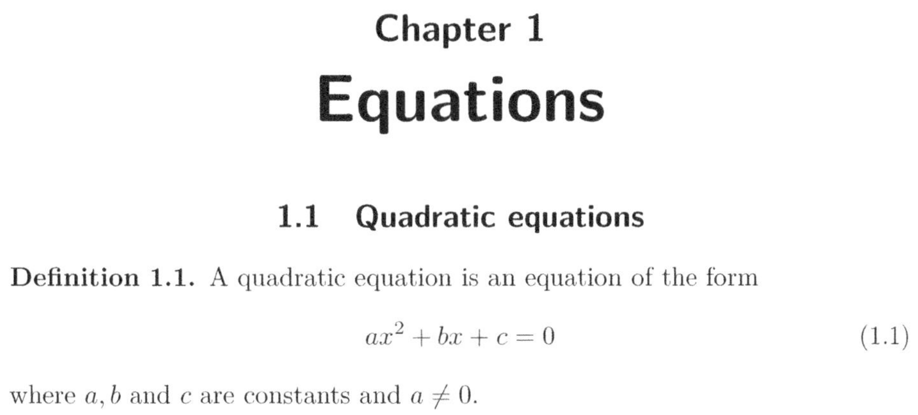

Chapter 1

Equations

1.1 Quadratic equations

Definition 1.1. A quadratic equation is an equation of the form

$$ax^2 + bx + c = 0 \tag{1.1}$$

where a, b and c are constants and $a \neq 0$.

Figure 12.6 – Centered headings

In *step 1*, we loaded the `titlesec` package, which provides a comprehensive interface for customizing headings of parts, chapters, sections, and even smaller sectioning parts down to subparagraphs.

In *step 2*, we selected a **display** style that places the heading number and the title on separate lines. We first switched to the base font using the `\normalfont` command, to be on the safe side. By applying `\sffamily`, we switched to a sans-serif font, set the size and weight, and finally, declared that the entire heading shall be centered.

In *step 3*, everything is very similar to *step 2*; we just omitted `[display]` to get the number and title on the same line.

To understand the remaining arguments, have a look at the general form of the`\titleformat` command:

```
\titleformat{cmd}[shape]{format}{label}{sep}{before}[after]
```

The meaning of the arguments is as follows:

- `cmd` stands for the sectioning command that we redefine, that is, `\part`, `\chapter`, `\section`, `\subsection`, `\subsubsection`, `\paragraph`, or `\subparagraph`.
- `shape` specifies the paragraph shape, with the effect of the possible values as follows:
  - `display` puts the label into a separate paragraph
  - `hang` creates a hanging label, like in standard sections, and is the default option
  - `runin` produces a run-in title like `\paragraph` does by default
  - `leftmargin` sets the title into the left margin

  - `rightmargin` puts the title into the right margin
  - `drop` wraps the text around the title, and requires care to avoid overlapping
  - `wrap` works like `drop`, but adjusts the space for the title to match the longest text line
  - `frame` works like `display` and additionally frames the title

- `format` may contain commands that will be applied to the label and text of the title.
- `label` prints the label, that is, the number.
- `sep` is a length that specifies the separation between the label and title text. With the `display` option, it's the vertical separation. With the `frame` option, it means the distance between the text and the frame. Otherwise, it's the horizontal separation between the label and title.
- `before` can contain code that comes before the title body. The last command is allowed to take an argument, which should then be the title text.
- `after` can contain code that comes after the title body.

That's a lot of options. Have a look at the `titlesec` documentation to learn even more by running `texdoc titlesec` or visiting `https://texdoc.org/pkg/titlesec`.

We used the `titlesec` command called `\chaptertitlename`, which defaults to `\chaptername`. So, it defaults to **Chapter**. In an appendix, it changes to `\appendixname`.

Using the following command, we customized the spacing of all chapter headings:

```
\titlespacing*{cmd}{left}{beforesep}{aftersep}[right]
```

The arguments have the following meanings:

- `left` works differently depending on the chosen `shape`: with `drop`, `leftmargin`, and `rightmargin`, it's the title width. With `wrap`, it's the maximum width. With `runin`, it sets the indentation before the title. Otherwise, it increases the left margin. If negative, it decreases, meaning it overhangs into the margin.
- `beforesep` sets the vertical space before the title.
- `aftersep` sets the separation between the title and text. With a `hang`, `block`, and `display` shape, it has a vertical meaning. With a `runin`, `drop`, `wrap`, `leftmargin`, and `rightmargin` shape, it's a horizontal width. Again, it may be a negative value.
- `right` increases the right margin when a `hang`, `block`, or `display` shape is used.

If you use the star, `titlesec` removes the indentation of the following paragraph. That's like standard sections: the text that follows a section heading doesn't have paragraph indentation. With `drop`, `wrap`, and `runin`, the starred version has no meaning.

In our example, we avoided indenting the paragraph following a chapter heading and specified a `30pt` space before the heading and a `20pt` space after it. That's less compared to standard classes, which use `50pt` above chapter headings.

It's highly recommended to read the `titlesec` documentation to get the most out of it. Its appendix shows how the headings in standard classes would be defined using `\titleformat` and `\titlesec`. That's a great way to start: copy these definitions and modify them.

Using sans-serif headings is very common today. They don't have the heavy, ancient appearance of bold serif headings. However, serif text offers the best readability for body text. Now, it's up to you to choose – you've got the tools.

Summary

In this chapter, we enhanced our document with a hypertext structure including colored links and bookmarks for navigation. We can now edit the PDF metadata and customize the styles of our headings.

In the next chapter, we will explore new ways of styles and navigation.

13

Creating Presentations

In the previous chapter, we used hyperlinks to navigate as a reader. When you make a presentation, you need navigation for the speaker. So far in this book, we've focused on documents meant to be read on paper. Presenting live, on projectors or screens, in academic talks or in **Zoom** or **Microsoft Teams** sessions, requires a different approach to visual design and navigation tools. Precisely for this purpose, LaTeX offers a dedicated presentation class with built-in support for slide design, navigation, and presentation, prioritizing speaking over reading. It's called the `beamer` class, named after the traditional term for a video projector that beams light onto a screen.

In this chapter, we'll use this class and work through the key topics needed to create presentations with LaTeX. We'll cover the following topics:

- Getting started with the `beamer` class
- Using overlays and incremental reveals
- Arranging columns
- Using blocks
- Generating a handout

Let's begin by setting up our first presentation.

Getting started with the beamer class

We'll build a small presentation step by step to see how the `beamer` class works in practice. It shall represent a fictional talk by **Till Tantau**, the creator of the `beamer` class, with thoughts on best practices.

Follow these steps in your editor:

1. Start the document using the `beamer` class:

```
\documentclass{beamer}
```

2. Choose a **theme** that defines the visual design. Here, we use the `Warsaw` theme, which provides a pretty rich and structured layout:

```
\usetheme{Warsaw}
```

3. Next, add some metadata for the presentation, such as the title, author, date, and affiliation:

```
\title{How to Give a Presentation}
\subtitle{And Keep Everyone Awake}
\author{Prof. Dr. Till Tantau}
\institute{Universität zu Lübeck}
\date{April 1, 2026}
```

4. Begin the document:

```
\begin{document}
```

5. From this point on, every slide lives inside a `frame` environment. We start with a title page:

```
\begin{frame}
  \titlepage
\end{frame}
```

6. Add a slide with a table of contents to give the audience an overview:

```
\begin{frame}
  \frametitle{Overview}
  \tableofcontents
\end{frame}
```

7. Declare a section and a subsection:

```
\section{Planning the talk}
\subsection{Things That Work}
```

8. Now add some actual content. Keep text short and focused. Lists work particularly well in presentations. Again, everything goes into a `frame` environment, and you can give it both a title and a subtitle:

```
\begin{frame}
  \frametitle{Favor the main message over details}
  \framesubtitle{Define the overall structure early}
  \begin{itemize}
  \item Plan around the available time
  \item Favor the main message over details
  \item Keep section titles self-explanatory
  \item Follow a clear logical flow
  \item Start by stating the topic and goal
  \item End with a short, clear summary
  \end{itemize}
\end{frame}
```

9. Add another subsection and frame:

```
\subsection{Things That Don't}
\begin{frame}
  \frametitle{What to Avoid}
  \framesubtitle{How to not mess it up}
  \begin{itemize}
  \item Don't read slide text out loud
  \item Don't talk too fast
  \item Never overload slides with content
  \item Avoid subsubsections
  \item Don't speak to the slides, talk to the audience
  \item Cut details, put them into the appendix
  \end{itemize}
\end{frame}
```

10. To get a better feeling for structure and navigation, add a few more sections:

```
\section{Structuring the Presentation}
\section{Building the Presentation}
```

11. Finish the document with one more frame so that the last section is not empty; otherwise, sections without following content would be ignored:

```
\begin{frame}
  More text
\end{frame}
\end{document}
```

12. Compile the document at least twice so LaTeX has the chance to generate the table of contents and navigation entries.

Look at the output. This is the first slide:

How to Give a Presentation

And Keep Everyone Awake

Prof. Dr. Till Tantau

Universität zu Lübeck

April 1, 2026

Figure 13.1 – The title slide

We got a well-designed title slide with all metadata included. You can see the section navigation at the top, and navigation symbols at the bottom. Let's zoom in on the navigation button bar:

Figure 13.2 – The navigation button bar

Clicking those interactive symbols allows you to move slides forward and backward and jump between sections. This works in any PDF viewer that supports links, such as Adobe Acrobat Reader.

This is how they work, from left to right:

- A click on the first rectangle, a slide icon, allows you to enter the slide number you want to jump to. The left and right arrow lets you jump a slide forward or back.
- The stacked rectangles stand for a frame with several slides in an overlay. Clicking the left arrow jumps to the beginning of the frame, the right arrow moves you to the end of the frame, skipping step-by-step uncovering on that slide.
- The arrows next to the subsection icon are used to navigate to the beginning or the end of a subsection.
- The next symbol looks like a highlighted section with subsections; here, the arrows move you to the first or last slide of a section.
- Then we have the table of contents symbol. Clicking on its left side jumps to the very beginning, and clicking on the right side moves you to the last slide. In case you have an appendix, another such symbol will appear that does the same for the appendix, jumping to the beginning or end slide.
- The magnifying glass symbol is the "search" or "find" icon, where you can enter a text string to search the whole presentation. The arrows beside it are used to move between previously visited slides.

Once you've made a small presentation, try this navigation. Of course, you can also use the standard navigation of your PDF reader, including the arrow keys. I admit that I usually take the easy way by just using the forward and backward arrow keys.

Note that you can remove the beamer navigation bar by disabling it in the preamble as follows:

```
\setbeamertemplate{navigation symbols}{}
```

The second slide is our table of contents that displays the sections and subsections:

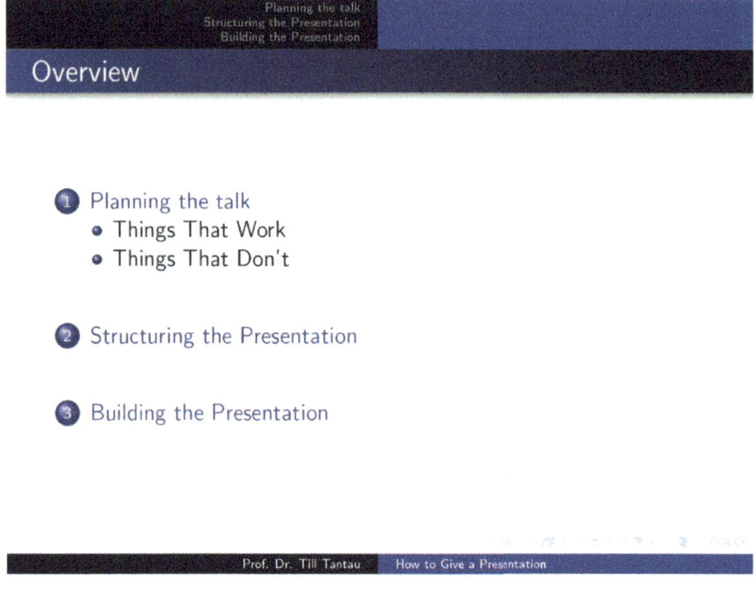

Figure 13.3 – The table of contents

The next slides are the actual content. Take a look, especially at the top of the slide:

- Plan around the available time
- Favor the main message over details
- Keep section titles self-explanatory
- Follow a clear logical flow
- Start by stating the topic and goal
- End with a short, clear summary

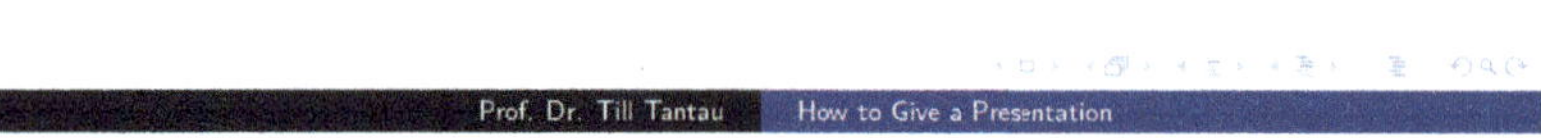

Figure 13.4 – A slide with content

At the top, you can see the current section highlighted on the left side and the current subsection highlighted on the right side. You can click the section or subsection title to move straight to that section or subsection.

As you go to the next slide, note the change in highlighting in the top-right area:

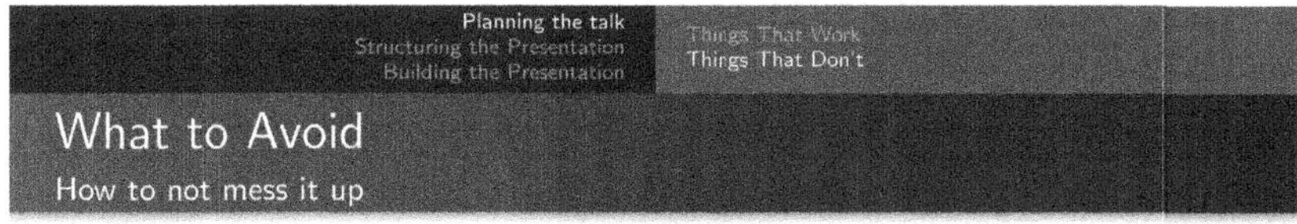

Figure 13.5 – A slide from the next subsection

You can show your presentation using any PDF reader on your laptop connected to a projector or on a shared screen in a Zoom or Teams session.

The default aspect ratio of `beamer` slides is 4:3. Many modern projectors and screens use 16:9, though. You can switch to that format by adding a document class option:

```
\documentclass[aspectratio=169]{beamer}
```

Note that the option doesn't use a colon (:). Beamer expects a numeric keyword, not a ratio written as 16:9. You set the aspect ratio by giving digits, and beamer interprets them based on how many there are:

- Two digits: `aspectratio=43` means 4:3
- Three digits: `aspectratio=128` means 12:8
- Four digits: `aspectratio=1610` means 16:10

In short, you pass the ratio as digits, and beamer does the math to compute the slide width, always in landscape format. Internally, the beamer class also chooses a suitable paper size so that font sizes look right onscreen.

Using overlays and incremental reveals

When I listen to a talk with slides, I often catch myself reading ahead. While the speaker is still explaining the first point, my eyes are already on the last bullet. This makes it harder to follow the spoken explanation, especially when the slide can be read faster than it can be explained. While a slide shows everything at once, you may often prefer parts of it to appear only when you start talking about them.

This is what **overlays** are for. One `frame` in the source file can produce several slides. Parts of the content can appear, disappear, or change from one slide to the next. Overlays control when something becomes visible, for example, when revealing list items step by step or building a formula in stages. We will start with the simplest overlays and then move on to more precise control.

Pausing content

The `\pause` command is the simplest way to create overlays. It lets a frame unfold step by step. Within a `frame`, everything before a `\pause` is shown on the current slide; everything after it is delayed until you advance to the next one. Each additional `\pause` adds another step. All of this happens within a single `frame` environment.

You can place `\pause` almost anywhere, including inside environments such as `itemize`, as shown here:

```
\begin{frame}
\begin{itemize}
  \item Keep one clear idea per slide
  \pause
  \item Use short phrases, not full sentences
  \pause
  \item Use large, readable fonts
\end{itemize}
\end{frame}
```

This produces three slides, each revealing one additional point, with the final slide showing all three.

Using incremental specifications

You can use angle brackets < > to provide **timing information**. They are rarely used in normal text, so they stand out clearly without getting in the way.

For example, `<2-3>` means that the content is visible on overlays 2 and 3 of the current `frame`. The numbers tell `beamer` when the content appears and until when it stays visible. We call this an **overlay specification**.

More practical for everyday use, the short syntax `<1->` simply means from overlay 1 onward. The number tells `beamer` when the content first appears, and the dash just means that it remains visible on all following overlays of the frame. So, instead of using the `\pause` command, you can write a list like this, working the same:

```
\begin{itemize}
  \item<1-> Use a consistent color scheme throughout the slides
  \item<2-> Ensure sufficient contrast for readability
  \item<3-> Keep visual elements balanced on the slide
\end{itemize}
```

Using shorthand

Writing explicit overlay numbers gives you full control, but it can become tedious for long lists. `beamer` can count overlays for you. If you use a + inside an overlay specification instead of a number, `beamer` automatically advances the overlay counter. Each + stands for *the next overlay*. There are these shortcuts:

- `\item<+->` produces a point for the next overlay that stays visible
- `\item<+>` shows a point the next overlay on, not seen in later overlays

That way, you no longer need to renumber items when you insert or remove lines.

You can make this even shorter: instead of adding an overlay specification to each `\item`, you can put it on the list environment itself:

```
\begin{itemize}[<+->]
```

That way, `beamer` applies `<+->` to every item automatically. Each item appears on the next overlay and stays visible afterward. This is usually the cleanest way to create incremental lists.

To show the table of contents one section at a time using overlays, use `\tableofcontents[pausesections]`. With `\tableofcontents[pausesubsections]`, the same applies at the subsection level.

Arranging columns

Some slides work better when content is placed side by side – for example, text next to a figure or two lists. With `beamer`, you can split a `frame` into vertical areas using **columns**. Each column holds its own content, while everything stays aligned. They are designed as LaTeX environments and help keep slides clear and structured.

Let's take a look at a simple example slide, using some content just for fun:

```
\begin{frame}
  \begin{columns}
    \begin{column}{0.55\textwidth}
      What I don't want to say:
      \begin{itemize}
        \item This slide looked better yesterday
        \item Just believe this
        \item I'll skip this quickly
        \item Time is running out
      \end{itemize}
    \end{column}
    \begin{column}{0.4\textwidth}
      \includegraphics[width=4cm]{ctanlion.pdf}
    \end{column}
  \end{columns}
\end{frame}
```

The `columns` environment groups multiple columns inside a frame. It does not take a width of its own. The width is given as the mandatory argument of each `column` environment. As shown here, it is convenient to specify it as a fraction of `\textwidth`.

This is how the slide looks:

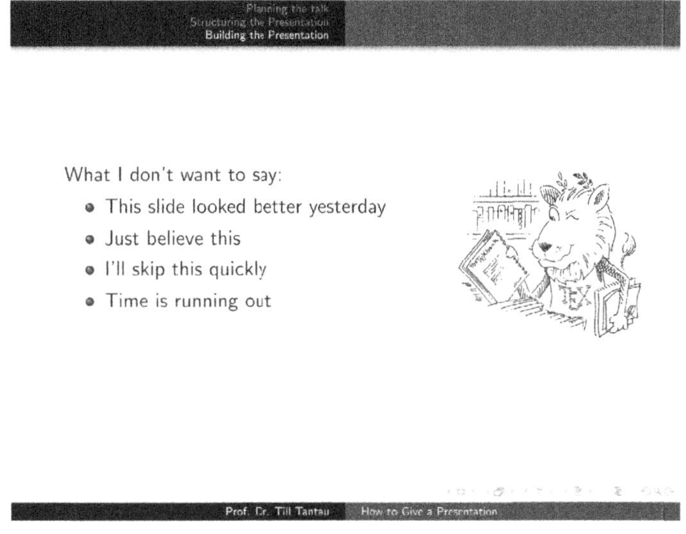

Figure 13.6 – A slide from the next subsection

Both environments can also take an optional alignment argument, like this:

```
\begin{columns}[t]
\begin{column}[t]{0.5\textwidth}
```

This controls how the columns are aligned vertically, and how the content inside the column is aligned relative to the other columns:

- `t`: top-aligned
- `c`: vertically centered
- `b`: bottom-aligned

Using blocks

To emphasize key points, you can use **blocks**. They separate an important idea from the surrounding text and are useful for definitions, key statements, examples, short conclusions, or warnings.

The `beamer` class provides three standard block types:

- `block` for neutral emphasis; for example:

  ```
  \begin{block}{Using blocks}
    You give the block a title and place the content inside it.
  \end{block}
  ```

- `exampleblock` for examples or demonstrations, like this:

```
\begin{exampleblock}{You see}
  That's how a block looks in a positive way.
\end{exampleblock}
```

- `alertblock` for important or warning-style content, like here:

```
\begin{alertblock}{Warning}
  Use blocks sparingly and intentionally. Not like on this slide.
\end{alertblock}
```

They all work the same and differ only in their visual style. With the `Warsaw` theme, the slide will look like this:

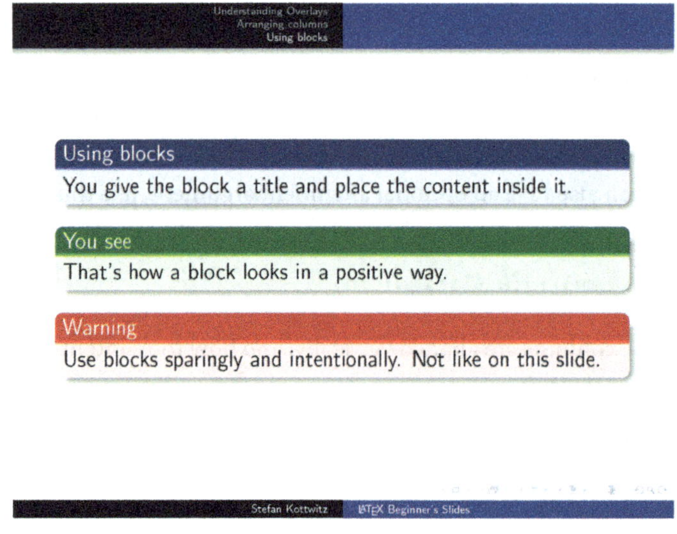

Figure 13.7 – Block types

With another theme, blocks can look different, as every theme has its own visual and structural style. Let's take another theme in the next section.

Generating a handout

After a talk, the audience may want something they can read at their own pace. That is what a handout is for: a static version of your presentation without overlays, often arranged to fit several slides on one page for printing.

The `beamer` class can generate such a handout directly from the same source file, by using overlays that are flattened for reading rather than for presenting. That's done quickly like this:

1. Add the `handout` option to the `document` class:

```
\documentclass[handout]{beamer}
```

2. Load the `pgfmorepages` package to place several slides on one page:

```
\usepackage{pgfmorepages}
```

3. Choose a layout – here, four slides on one page, slightly shrunk:

```
\pgfpagesuselayout{4 on 1}[a4paper, border shrink=0.25cm, landscape]
```

4. Let's use a different theme so you see how much the look and structure can change. We will get the section navigation and meta information on the left side instead of at the bottom, and square bullets instead of round bullets:

```
\usetheme{Berkeley}
```

5. Compile, and have a look at the PDF file. This is the first page, A4 paper in landscape:

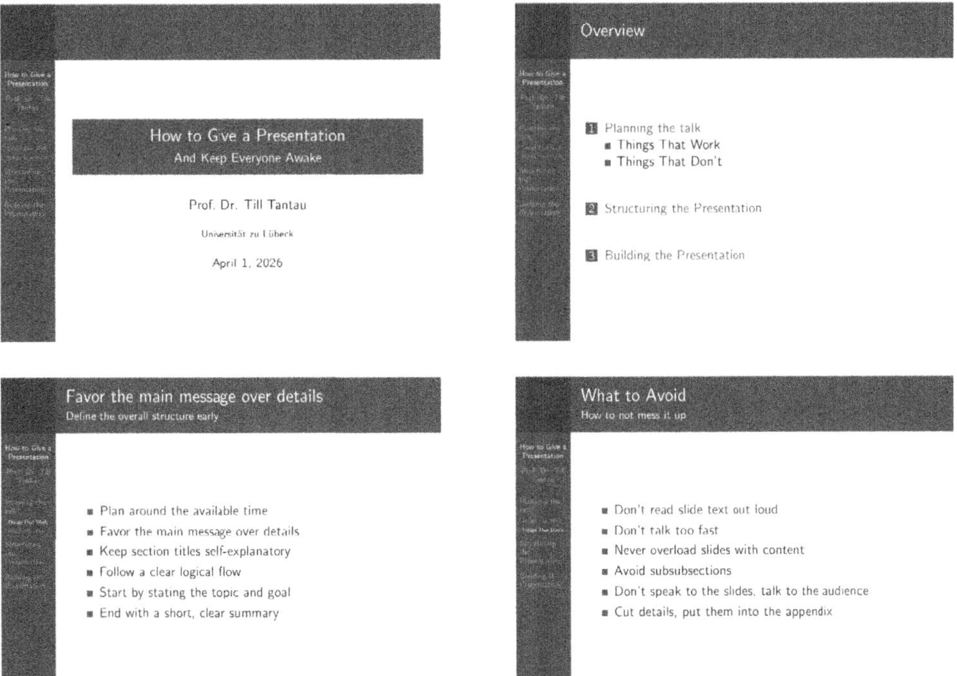

Figure 13.8 – A handout for printing

You can have up to 16 slides on one page, and different arrangements. For more information, visit `https://texdoc.org/pkg/pgfmorepages`.

Summary

This chapter gave a fast-paced introduction to building your own presentation, using as little technical syntax as possible. The `beamer` class is very powerful and comes with many features and details to explore. If you want to go further, have a look at the manual at `https://texdoc.org/pkg/beamer` or run `texdoc beamer` at the command line.

The *LaTeX Cookbook* shows some examples and some handy solutions in *Chapter 1, Exploring Various Document Classes*. You can find the code of the examples with screenshots at `https://latex-cookbook.net/chapter-01`.

To explore the variety of `beamer` themes, visit `https://latex-beamer.net`.

If you want to check out a newer, still experimental presentation class, visit `https://texdoc.org/pkg/ltx-talk`. It's less complex and focuses on accessibility with modern PDF tagging.

During our work, we may encounter errors and warnings. That's common for advanced LaTeX users as well. The following chapter will prepare us for troubleshooting.

14

Troubleshooting

Now that we have built complete documents and entire presentations, it's time to look at what happens when things don't work as expected. Sooner or later, every LaTeX project runs into errors or warnings. That's perfectly normal and is often caused by minor issues, such as typos in command names or unbalanced braces. Even professional LaTeX users deal with errors daily; they know how to handle them efficiently. Let's learn how to do that too.

In this chapter, we'll look at how to deal with common problems in LaTeX, including the following:

- Understanding and fixing errors
- Handling warnings
- Avoiding obsolete classes and packages
- General troubleshooting

Let's start by looking at how to resolve errors.

Understanding and fixing errors

When the LaTeX engine runs into a problem, it reports an error. These messages are meant to help, so it's worth reading them carefully. Along with the line number where the problem occurred, LaTeX provides a short diagnostic message describing what went wrong.

Focus on the very first error message you see. If you continue compiling, any following errors are often just side effects of the original problem that confused the compiler.

Let's create a small test document. You may know the classic "**Hello world!**" example from learning programming languages—we'll do the same in LaTeX. Though we are familiar with the special capitalization used in the words TeX and LaTeX, let's see what happens if we try to use the `\Latex` command instead:

1. Create a new document with these lines:

```
\documentclass{article}
\begin{document}
\Latex\ says: Hello world!
\end{document}
```

2. Compile the document. LaTeX will stop and report this error message:

```
! Undefined control sequence.
l.3 \Latex
\ says: Hello world!
```

3. Click the **Cancel** button to stop compiling; in TeXworks, it's the **Cancel** icon in the upper-left corner.
4. In our code from *step 1*, go to line 3, and replace `\Latex` with `\LaTeX`. Then compile again. This time, LaTeX produces output without any errors:

LaTeX says: Hello world!

Figure 14.1 – The output of the corrected document

LaTeX commands are **case-sensitive**. Since we didn't respect that rule, LaTeX had to deal with a macro called `\Latex`, which it simply doesn't know. Because a LaTeX command is also called a **control sequence**, the error message reports an **Undefined control sequence**.

When LaTeX encounters an error, it stops processing and waits for user input. You could press the *Enter* key to continue, though, but this often results in incomplete or incorrect output. In most cases, it's better to cancel the compilation and correct the error right away.

Let's break down the error message into its main parts:

- An error message starts with an exclamation mark, followed by a short description of the problem

- Next, LaTeX indicates the input line number where the error occurred and highlights the portion of that line that caused it
- After a line break, LaTeX prints the remaining part of the input line for context

So, you're not left guessing. LaTeX tells you precisely what you need to know:

- The type of error
- The precise location where it happened

Most editors either display the line number or allow you to jump to the line number you enter. As you can now easily find the problematic spot in the source code, you only need to know why LaTeX is complaining—that's what we'll look at in the following sections.

If you're using Overleaf, there's one small caveat: Overleaf hides the error messages, continues compiling, and shows output even when an error occurs. The following screenshot shows how our document looks in Overleaf with the error present:

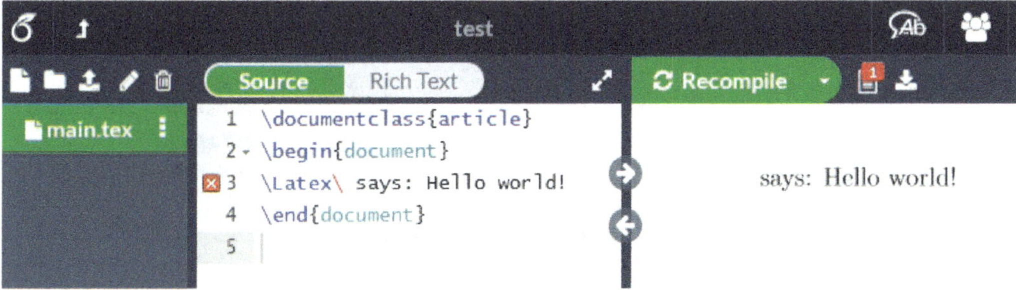

Figure 14.2 – A code error seen in Overleaf

At first glance, our document appears to compile. But on closer inspection, we notice the following:

- At the beginning, the word `LaTeX` is missing
- A small red number appears above the output

That small red number indicates an error and should not be ignored. Otherwise, as we saw here, we may get a document with missing or incorrect output, and in a large document, that might be hard to notice.

Click the red number to open a window with the error message:

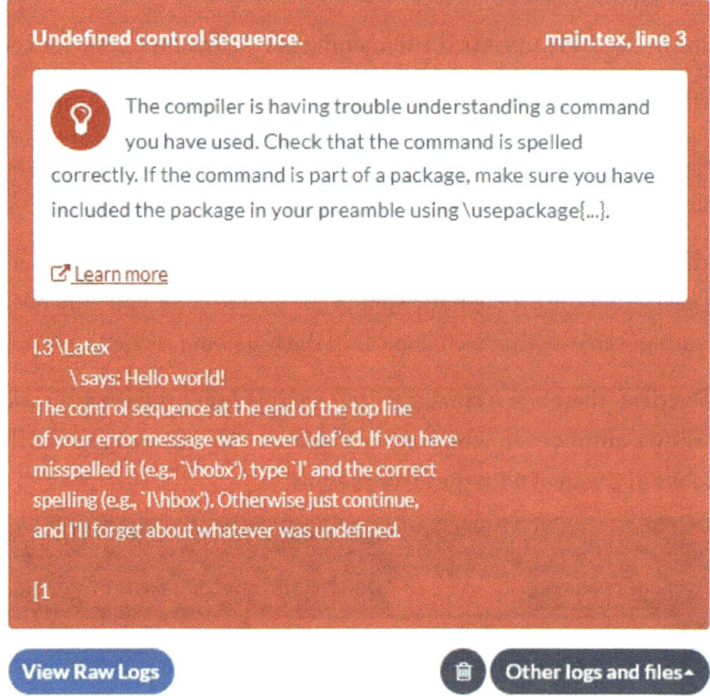

Figure 14.3 – An error message in Overleaf

Now we can see the full error message, the explanation, and the exact location at **line 3**. In *Figure 14.2*, the problematic line 3 is marked with a red cross. You can view the complete log file with detailed information, errors, and warnings by clicking **View Raw Logs** in the lower-left corner.

Next, we'll take a closer look at common TeX and LaTeX error messages. We'll go through them step by step in the following sections, starting with errors that occur in the preamble.

Handling the preamble and document body

The preamble contains all document-wide settings. This is where we choose the document class, load packages, set options, and define commands. The `\begin{document}` command marks the end of the preamble and the start of the document body, where the actual content goes. If something goes wrong with this structure, LaTeX will typically report one of the following errors:

- **Missing \begin{document}**: In many cases, that simply means that the `\begin{document}` command was forgotten. However, the error can also appear if the command is present. A common reason is that a character or command in the preamble produces output; you can spot and remove it. Just remember that output is not allowed before `\begin{document}`.
- **Can be used only in preamble**: This error message indicates that a command that is meant to be used in the preamble was used after `\begin{document}`. A typical example is the `\usepackage` command. Move the command back up into your preamble or remove it if it doesn't belong there.
- **Option clash for package**: An option clash occurs when LaTeX loads a package more than once but with different options. That can happen if you have two `\usepackage{...}` lines for the same package in your document preamble. If you did that, it's usually better to reduce it to one `\usepackage` call with the desired options. Sometimes the cause is less obvious: a class or a package may already load the package along with some options. If you want to load the package too, but with different options, LaTeX complains.

You could resolve an option clash by omitting the package reload and specifying the desired options in the document class options. Remember, packages inherit class options. Some packages and classes even offer commands to set options after loading. For example, the `hyperref` package provides `\hypersetup{options}`, and similarly, the `caption` package offers `\captionsetup`.

In the following sections, we will look at common issues in the document body.

Using commands and environments

Command names might be easily misspelled or misused. Let's check out LaTeX's common complaints:

- **Undefined control sequence**: As in our example in the previous section, LaTeX stumbled across an unknown command name. There are two possible reasons:
  - The command name might be misspelled. In that case, you just need to correct it and restart the typesetting.
  - The command name is correct, but it's defined by a package you didn't load. Add a `\usepackage` command to your preamble to load the required package.
- **Environment undefined**: This is similar to **Undefined control sequence**, but this time, we began an unknown environment. Again, this may be caused by a misspelling or by a missing package—you know how to correct it.

- **Command already defined**: This happens when you create a command with a name that's already used, for example, with `\newcommand` or `\newenvironment`. Just choose a different name. If you would really like to override that command, use `\renewcommand` or `\renewenvironment` instead, but be careful with redefining an existing command: if it's already used internally by LaTeX or packages, changing it can cause side effects.
- **Missing control sequence inserted**: A control sequence was expected but didn't appear. A common cause is using `\newcommand`, `\renewcommand`, or `\providecommand` without specifying a command name as the first argument.
- **\verb illegal in command argument**: The `\verb` command for producing verbatim text is a delicate one; it cannot be used within arguments of commands or environments. The `examplep` package offers commands for using verbatim text in such places.

Of this list, the very first error is probably the one that happens most often, since typing errors happen as easily as forgetting to load a package.

Writing math formulas

When LaTeX encounters an error during typesetting math expressions, one of the following error messages can occur:

- **Missing $ inserted**: Many commands may only be used in math mode. Just think of symbols; most of them require math mode. If LaTeX is not in math mode and encounters such a symbol, it stops and prints that error. Usually, we can resolve such errors by inserting that missing `$`. Forgetting to start or end math mode is one of the most frequent mistakes. Also, remember that you cannot use paragraph breaks inside a math expression. This means that blank lines within a math expression are illegal; we have to end math mode before the blank line.
- **Command invalid in math mode**: Some commands are not applicable within math formulas. In that case, use the command outside math mode.
- **Double subscript, double superscript**: Two subsequent subscripts or superscripts cannot be compiled. For example, in `$a_n_1$`, LaTeX cannot decide if `a_n` should have a subscript `1`, or if `a` should have a subscript `n_1`. To correct that, group them by braces, such as in `$a_{n_1}$`.
- **Bad math environment delimiter**: This can result from illegally nesting math mode. You must not start math mode if you are already inside math mode. For example, don't use `\[` within an `equation` environment. Similarly, you must not end math mode before you start it. Ensure that your math mode delimiters match and that braces are balanced.

In *Chapter 9, Writing Math Formulas*, we learned how to avoid such errors.

Working with files

If LaTeX cannot open a file for you, it can raise one of the following errors:

- **File not found**: LaTeX tried to open a nonexistent file. Possibly, you did one of the following:
    - Used `\include` or `\input` to include a `.tex` file, but a file with the specified name doesn't exist
    - The file has spaces or special characters in the name; better avoid that
    - Tried to use a nonexistent package or misspelled the package's name
    - Used a document class that doesn't exist or has a different name

  Just correct the filename in your input document or rename the file.
- **\include cannot be nested**: We learned in *Chapter 11, Developing Large Documents,* that we cannot use the `\include` command within a file that is being included itself. Instead, we use `\input` within such files.

It's good to avoid special characters and spaces in filenames. Both LaTeX and the operating system may have issues with unusual characters in filenames, so it's good to stick to common letters, digits, dashes, and underscores.

Creating tables and arrays

Admittedly, `tabular` and `array` environments don't have the most straightforward syntax. Those `&` and `\\` might easily be misplaced, which causes LaTeX to complain. Further, we need to be careful with the formatting of arguments. These are the possible errors you might see regarding arguments to the environment:

- **Illegal character in array arg**: In the argument to a `tabular` or an `array` environment, you can specify the column formatting. You line up characters such as `l`, `c`, `r`, `p`, `@`, and width arguments such as `{1cm}`. If you use any character that doesn't have such a meaning, LaTeX will tell you. The same applies to the formatting argument of `\multicolumn`.
- **Missing p-arg in array arg**: A bit more specific than the previous message, this tells us that the width argument to the `p` option is missing. Supplement `p` with a width such as `{1cm}` or change `p` to another option, such as `l`, `c`, or `r`.

- **Missing @-exp in array arg**: The expression after the `@` option is missing. You just need to add it, in curly braces, or remove the `@` option.

Now we shall take a look at the potential error messages concerning the table body:

- **Misplaced alignment tab character &**: As you know, the ampersand character has the special meaning of separating columns in a row of a tabular or array environment. If you accidentally use it in regular text, this error will appear. Type `\&` if you desire an ampersand symbol in the output.
- **Extra alignment tab has been changed to \cr**: This happens if you use too many `&` alignment tab characters, which are for dividing columns. For example, with two columns, we cannot have four `&` characters as column dividers. Such an error can happen if we forget to add `\\`, which ends a row.

In *Chapter 5*, *Creating Tables*, we discussed the proper syntax to avoid such errors.

Working with lists

Lists follow a specific structure and cannot be endlessly nested. At some point, LaTeX may complain, such as in the following error messages:

- **Too deeply nested**: We can nest up to four levels of a list. If we mix list types, we can go up to six levels. But if we go further than what LaTeX accepts, we will get this error message. Think about it if you really need deep nesting. If this is the case, consider using sectioning commands such as `\paragraph` or `\subsubsection` for outer levels.
- **Something's wrong--perhaps a missing \item**: An `\item` command is missing. We may have simple text in an `itemize` or `enumerate` list. Then we need to insert an `\item` command before that text.

In *Chapter 4*, *Creating Lists*, we learned the proper list syntax.

Working with floating figures and tables

In *Chapter 5*, *Creating Tables*, and *Chapter 6*, *Including Images*, we learned how to insert figures and tables and adjust their placement. If you use a lot of floating objects, that is, figures or tables, you might encounter this error: **Too many unprocessed floats**.

If you use a floating object and LaTeX doesn't find an appropriate place (e.g., there might be no space), it saves the object for later placement. If that happens often, LaTeX's room for floating objects may fill up, causing this error. It may be solved as follows:

- By adding placement options, such as `[htbp!]`, to the `figure` and `table` environments, thus lowering their placement requirements
- By inserting `\clearpage` to flush out the floats at a suitable place, or perhaps even cleverer: `\afterpage{\clearpage}` with the `afterpage` package

In the final section, we examine other potential error scenarios.

General syntax errors

Just as with any markup or programming language, LaTeX's documents have to follow a syntax. For example, braces and delimiters have to match. If there's a mistake, LaTeX will point to it:

- **Missing { inserted, missing } inserted**: Though it reads as unbalanced braces might cause it, it may be due to TeX being confused. Most likely, the error occurred before the point where LaTeX reports it. So, check the syntax used thoroughly.
- **Extra }, or forgotten $**: This time, there's a problem with unbalanced braces, or math mode delimiters don't match correctly. You need to correct the matching.
- **There's no line here to end**: Using `\\` or `\newline` between paragraphs in vertical mode is not meaningful and causes this error. Don't try to get more vertical space by writing `\\`. Use `\vspace` instead, or other skip commands such as `\bigskip`, `\medskip`, or `\smallskip`. For instance, we can produce a blank line with `\vspace{\baselineskip}`.

The frequently asked questions list for TeX and LaTeX, called **TeX FAQ**, lists error messages together with explanations and suggestions. It's available at `https://texfaq.org/#errors`.

Once we fix all errors, there may still be some flaws in the document. LaTeX prints out warnings if it sees a potential issue. In the next section, we will see how to deal with them.

Handling warnings

Warning messages are for your information. They don't always point to a severe problem, but often it's a good idea to read these tips carefully and act accordingly. This may improve your document.

We will test this now. Let's say we want to emphasize text that is in a sans-serif font. We expect italic sans-serif text as a result.

Let's try this:

1. Take our "`Hello world!`" example and modify it this way:

```
\documentclass{article}
\renewcommand{\familydefault}{\sfdefault}
\begin{document}
\emph{Hello world!}
\end{document}
```

2. Compile. LaTeX will print out a warning in the log file:

```
LaTeX Font Warning: Font shape `OT1/cmss/m/it' in size <10> not
available
(Font) Font shape `OT1/cmss/m/sl' tried instead on input line 4.
```

3. Check the output:

Hello world!

Figure 14.4 – Slanted shape instead of italic shape

The `\familydefault` macro stands for the default font family used in the LaTeX document. For this macro, we specified the `\sfdefault` value, which is the default sans-serif font. This means sans-serif is now the default, regardless of the chosen font. As you can imagine, other possible values are `\rmdefault` and `\ttdefault`. By changing `\familydefault`, we don't have to write `\sffamily` repeatedly.

But then we emphasized our text and got a warning. The message means that there's no **Computer Modern Sans Serif (CMSS)** font in the default `OT1` font encoding, in medium weight (`m`) and italic shape (`it`) in `10pt` size. Furthermore, LaTeX told us how it tried to repair the problem – instead of italics, it chose a slanted shape. That's not too bad – at least it looks similar, and the output is produced.

This is basically what happens when warnings occur: LaTeX informs us of a potential problem or disadvantage, then tries to choose the best alternative and continues typesetting. It's not uncommon for a longer document to produce dozens of warnings, most often related to horizontal or vertical justification.

Often, it doesn't hurt if you ignore warnings that don't seem very serious, though following them up is a good habit. Anyone who desires to have a perfect document fixes all warnings. This way, we cannot overlook a potential problem.

In the following sections, we will deal with frequently occurring situations with warnings.

Justifying text

By default, LaTeX aligns the text at both the left and right margins. LaTeX does this by adjusting the space between words and letters. That is called **full justification.**

If LaTeX cannot achieve that, we may get one of the following warnings:

- **Overfull \hbox**: A line is too long and doesn't fit the text width. This may cause text to extend past the margin. This may be caused by hyphenation problems, which we can fix by using `\hyphenation` or by inserting `\-`, as you learned in *Chapter 2, Formatting Text and Creating Macros*. You could break the line manually or polish your words otherwise.
- **Underfull \hbox**: The opposite of the previous warning; a line is not wide enough to fit the text width, so LaTeX could not achieve full justification. This could be caused by `\linebreak`, if there's not enough text on the line. Also, `\\` or `\newline` may cause it, such as `\\\\`, since the text justification has been interrupted.
- **Overfull \vbox**: The page is too long because TeX could not break it accordingly. The text might hang out past the bottom margin.
- **Underfull \vbox**: There's not enough text on the page. TeX had to break the page too early.

In *Chapter 2, Formatting Text and Creating Macros*, we learned how to improve the justification, reducing such warnings. Remember, already loading the `microtype` package may help.

The `\sloppy` declaration switches to a pretty relaxed type of type-setting, thus avoiding many such warnings. Its counterpart is `\fussy`, switching back to the default behavior. Suppose you ever want to use `\sloppy` because a relaxed typesetting with possibly more stretched spacing is okay for you, then it's better to keep it local by grouping or by using the respective environment—`\begin{sloppypar}` ... `\end{sloppypar}`.

Additionally, check out recommendations and alternatives regarding `\sloppy` in `l2tabu`, mentioned in *Chapter 11, Developing Large Documents*.

Referencing

Many warnings deal with referencing. Common mistakes include missing labels or cite keys, duplicate keys, or just needing another typeset run.

The following warnings can occur:

- **Label multiply defined**: `\label` or `\bibitem` has been used with a label name that's already been used. Make label names unique; remember what we did in *Chapter 7, Using Cross-References*.
- **There were multiply-defined labels**: Like the previous warning, but after processing the complete document, two `\label` commands have defined the same label.
- **Labels may have changed. Rerun to get cross-references right**: Just typeset again to let LaTeX correct the referencing.
- **Reference ... on page ... undefined**: `\ref` or `\pageref` has been used without a corresponding `\label` definition. Insert a `\label` command at a suitable place.
- **Citation ... on page ... undefined**: A `\cite` command did not have a corresponding `\bibitem` command, or no BibTeX key in the `.bib` file.
- **There were undefined references or citations**: Summarizing after processing—any `\ref` or `\cite` command did not have a corresponding `\label` or `\bibitem` command.

Whenever you get warnings regarding referencing, it's a good idea to rerun typesetting. Often, such warnings then disappear because LaTeX couldn't resolve all references during the first run.

Choosing fonts

When LaTeX cannot use a font as needed, it may print one of the following warnings:

- **Font shape ... in size <...> not available**: You chose a font that's not available. This may be the result of combining font commands that produce a nonexistent font. Also, it could just be unavailable in that size. LaTeX will choose a different font or size and inform you about that choice in detail.
- **Some font shapes were not available, defaults substituted**: LaTeX prints this after processing the entire document if any of the chosen fonts were unavailable.

Check where such a warning occurs to see if the font size and shape are okay for you. Otherwise, you may consider using another font, as we did in *Chapter 10, Using Fonts*.

Placing figures and tables

Even if there are no errors, LaTeX may not be able to place a figure or table properly. In such a case, LaTeX may show one of the following warnings:

- **Float too large for page**: A figure or table is too large to fit the page. It would be printed, but the page would become too large.
- **h float specifier changed to ht**: If you specified an `h` option to a floating figure or table that doesn't fit there, it would be placed at the top of the next page, and that warning would be issued. The same can occur for `!h` and `!ht`.

Using all available placement options, such as in `\begin{figure}[!htbp]` or `\begin{table}[!htbp]`, as mentioned in *Chapter 6, Including Images*, can avoid many placement issues.

Customizing the document class

LaTeX may issue a warning reading **Unused global option(s)** if you use an illegal class option. This means you specified an option to `\documentclass` that's unknown to the class and to any loaded package. This could, for example, be a base font size that is not supported. Just check the option that LaTeX complains about.

Also, packages themselves may print out warnings if they foresee problems. All these warnings are intended to help you design your document, so it's good to look at each one.

Even if you get a document with no errors or warnings, it may still be imperfect if you use packages or classes that are no longer updated. We will look at some well-known obsolete packages in the next section.

Avoiding obsolete classes and packages

At the end of *Chapter 11, Developing Large Documents*, we discussed the dangers of outdated information. LaTeX has existed for decades, and so have tutorials, examples, packages, and templates. Many are totally obsolete, and some even refer to the old LaTeX standard 2.09, when document classes didn't exist. We pointed to the definitive guide, `l2tabu`, which comes to the rescue.

Many problems arise from the use of obsolete packages. For example, some that aren't maintained anymore may conflict with newer packages. Often, you need to find the recommended successor of an obsolete package and use that.

To help you in that matter, here's a short list showing obsolete packages and their respective recommended successors:

Obsolete packages	Recommended successors
a4, a4wide, anysize	geometry, typearea
backrefx	backref
bitfield	bytefield
caption2	caption
dinat	natdin
doublespace	setspace
dropping	lettrine
eps, epsfig	graphicx
euler	eulervm
eurotex	inputenx
fancyheadings	fancyhdr
floatfig	floatflt
glossary	glossaries
here	float
isolatin, isolatin1	inputenc
mathpple	mathpazo
mathptm	mathptmx
nthm	ntheorem
palatino	mathpazo
picinpar	floatflt, picins, wrapfig
prosper, HA-prosper	powerdot, beamer
ps4pdf	pst-pdf
raggedr	ragged2e
scrlettr	scrlttr2
scrpage, scrpage2	scrpage-scrlayer
seminar	powerdot, beamer
subfigure	subfig, subcaption
t1enc	fontenc
times	mathptmx
utopia	fourier
vmargin	geometry, typearea

Figure 14.5 – Obsolete packages and recommended successors

That's not set in stone. Of course, you may still use the so-called obsolete packages. They may work well even today. But check out their description on their CTAN package home page. Usually, there are comments on whether the packages are still relevant or obsolete, and it also lists recommended alternative packages. You can visit the package homepage at the URL starting with `https://ctan.org/pkg/`, followed by the package name, such as `https://ctan.org/pkg/geometry` for the `geometry` package.

You can find an updated version of that list at `https://latexguide.org/obsolete`.

Further, we will continue with some general advice in the next section.

General troubleshooting

There may be situations where we cannot solve a problem simply by reading and acting on warnings or error messages. Imagine a mysterious error, an untraceable error location, irresolvable references, or just unclear messages from classes or packages.

Locating the cause by the line number printed out by LaTeX, or by knowing what we've done since the previous typesetting run, usually helps. Once we've found a problematic line or chunk, we can remove or fix it. Otherwise, it might become difficult.

Here are the first general steps we can work through:

- **Compile several times**: This may be necessary for correct referencing, proper positioning of floating figures, and the creation of a table of contents, bibliographies, and lists of tables and figures.
- **Check the order in which you load the packages**: Some packages, such as `hyperref`, don't work well if loaded before or after specific packages. You may swap some lines to correct or test that.
- **Remove auxiliary files**: If anything goes wrong, it's sometimes a good idea to remove all files generated by LaTeX during the compilation process. These files have the same name as the main document but with extensions such as `.aux`, `.toc`, `.lot`, `.lof`, `.bbl`, `.idx`, or `.nav`, among others.

If the problem persists, we could try to isolate the cause as follows:

1. Create a copy of your document. If necessary, copy the complete folder. From now on, work on the copy.
2. Remove parts of the document that are likely not relevant to the problem. You can comment out lines by adding a `%` symbol in front.

3. Typeset to ensure that the problem persists. If so, go back to *step 2* and remove another part of the document. If the problem is gone, you have isolated it to the part you just removed. In this latter case, restore the deleted part since it contains the issue. If that part of the document is still too large to pinpoint the error, you can go back to *step 2* and try again by removing smaller parts.
4. After some repetitions of this process, you will have located the problem. If you didn't find it, reduce the number of loaded packages and repeat *steps 2* and *3*.
5. You will end up with a small but complete example document that reproduces the error. We call this a **minimal working example** (MWE).

Removing or rewriting that identified part of your document could help. What if you really want to use that part and would like to fix that error? Now that you can show the problem with a short code example, you could post that problem to an online LaTeX forum and ask for help.

You are not dependent on just the errors and warnings that your editor shows you. LaTeX keeps track of all information, each warning, and every error. These will be collected in a file named after your main document, with the `.log` extension. This is an ordinary text file that we can open in any editor, including your LaTeX editor.

For instance, the log file for our `Hello world` example at the beginning of this chapter starts with information about the TeX and LaTeX format versions and looks like the following:

```
This is pdfTeX, Version 3.141592653-2.6-1.40.22 (TeX Live
2021) (preloaded format=pdflatex 2021.6.25)  12 JUL 2021
00:47
entering extended mode
restricted \write18 enabled.
%&-line parsing enabled.
**document
(./document.tex
LaTeX2e <2021-06-01> patch level 1
L3 programming layer <2021-06-18>
```

It continues with information about the document class, its version, and the used class options `.clo` file:

```
(/usr/local/texlive/2021/texmf-dist/tex/latex/base
/article.cls
Document Class: article 2021/02/12 v1.4n Standard LaTeX
```

```
document class
(/usr/local/texlive/2021/texmf-dist/tex/latex/base
/size10.clo
File: size10.clo 2021/02/12 v1.4n Standard LaTeX file (size option)
)
```

It then shows the loaded packages and definitions – not many, in our case:

```
(/usr/local/texlive/2021/texmf-dist/tex/latex/l3backend
/l3backend-pdftex.def
File: l3backend-pdftex.def 2021-05-07 L3 backend support:
PDF output (pdfTeX)
\l__color_backend_stack_int=\count190
\l__pdf_internal_box=\box50
)
```

It tells us when it uses or opens a file:

```
No file document.aux.
\openout1 = `document.aux'.
```

It provides us with font information:

```
LaTeX Font Info:    Checking defaults for OML/cmm/m/it on input line 2.
LaTeX Font Info:    ... okay on input line 2.
```

It contains all errors and warnings:

```
! Undefined control sequence.
l.3 \Latex
          \ says: Hello world!
?
! Emergency stop.
```

Once we correct the error in *step 4* of the *Understanding and fixing errors* section, LaTeX adds information about LaTeX performance and memory to the log file:

```
Here is how much of TeX's memory you used:
385 strings out of 478510
6981 string characters out of 5849585
301299 words of memory out of 5000000
18443 multiletter control sequences out of 15000+600000
```

```
403430 words of font info for 27 fonts, out of 8000000
for 9000
1141 hyphenation exceptions out of 8191
34i,5n,41p,139b,107s stack positions out of
5000i,500n,10000p,200000b,80000s
</usr/local/texlive/2021/texmf-dist/fonts/type1/public/
amsfonts/cm/cmr10.pfb></usr/local/texlive/2021
/texmf-dist/fonts/type1/public/amsfonts/cm/cmr7.pfb>
```

The log file finishes by stating the output size as well as some statistics:

```
Output written on document.pdf (1 page, 22454 bytes).
PDF statistics:
18 PDF objects out of 1000 (max. 8388607)
10 compressed objects within 1 object stream
0 named destinations out of 1000 (max. 500000)
1 words of extra memory for PDF output out of 10000
(max. 10000000)
```

Check out the log files of some of the documents you've produced up to now. The information therein looks very technical, but this might help you a lot in troubleshooting.

Summary

This chapter prepared us to solve problems that might occur in our LaTeX document.

Specifically, we learned how to locate and fix errors, understand warning messages, and analyze LaTeX's typesetting log file.

Correcting errors is necessary. Remember to check `https://texfaq.org/#errors` for help. Dealing with warnings is a valuable bonus. If you encounter any problem that you cannot solve on your own, don't hesitate to ask for help on a LaTeX internet forum such as `https://latex.org`. In that forum, we have a section dedicated to this book, *LaTeX Beginner's Guide*, where I would be happy to answer your questions.

For LaTeX friends who are online, it's often an easy task to use this information to solve your problem. Definitely, a lot of LaTeX enthusiasts have fun helping other LaTeX users. The next chapter will discuss LaTeX forums on the internet and other online resources.

15

Using Online Resources

In the previous chapter, you mostly handled troubleshooting on your own. But you don't have to do that alone anymore. You can ask others online and work through problems together, for example, in internet forums and on community websites.

Over the years, a large body of LaTeX-related information has accumulated on the internet. Thanks to LaTeX's open source nature, a strong and active community has formed, sharing experience, solutions, and best practices.

In this chapter, we'll take a guided tour through the most useful online resources, including the following:

- Web forums, Q&A sites, and discussion boards
- Lists of frequently asked questions
- Mailing lists
- TeX user group websites
- Websites for LaTeX software and editors
- Graphics galleries
- LaTeX blogs
- X (formerly Twitter)

Many of the websites listed here are maintained by me and hosted on servers supported financially by DANTE e.V., the German-language TeX user group. You can find a complete list of my websites at `https://latex.net/about`.

Since you already know how to navigate the web, this chapter focuses less on hands-on examples and more on orientation. Let's start our tour with interactive discussion platforms.

Web forums, Q&A sites, and discussion boards

Let's start where most online interaction takes place: forums.

Internet forums, or web forums, provide an accessible and user-friendly way to ask questions, discuss problems, and share solutions. In the early days, LaTeX was discussed in subforums of general computer forums, alongside other software topics. As LaTeX grew in popularity, dedicated LaTeX-focused websites emerged, offering more specialized support. We'll look at some of these platforms in the following sections.

LaTeX.org

Launched in January 2007, `https://latex.org` was the first **web forum** dedicated to LaTeX. It's split into various subforums, each dealing with a particular LaTeX topic, such as *Math and Science* or *Fonts and Character Sets*, or a specific LaTeX distribution or editor.

Participating is as easy as in any other web forum. You don't need to register for reading, as it's freely available. Just for writing, you need to register once, choosing a login name and a password. Then, you may ask questions or support other users who went there looking for help.

Questions are very welcome! They are the foundation of the site. You may increase the chance of receiving helpful answers by doing the following:

- Choosing a meaningful header to get users interested in reading your question
- Describing your problem clearly
- Quoting the error or warning messages you've got
- Including a code example, which allows others to reproduce the problem

The latter is an excellent approach; there's even a website explaining why and how at `https://texfaq.org/FAQ-minxampl`. Once a problem can be reproduced, it's close to being solved, even if it seems complicated at first sight. Experienced users familiar with the source code of the LaTeX kernel and packages can explain how something works and can create solutions for nearly any problem. You provide problem code; you get solution code.

There is a subforum dedicated to this book, *LaTeX Beginner's Guide*: `https://latex.org/forum/viewforum.php?f=66`. There, you can post questions or remarks on this book, and I can answer them.

TeX and LaTeX on Stack Exchange

The website `https://tex.stackexchange.com` is a **question-and-answer (Q&A)** site that differs from classical web forums. While web forums are places where people talk and discuss, this Q&A site has a more straightforward structure. There's a question, followed by answers. There's no discussion except in the comments.

When you post a question, follow the same advice regarding posting as with `LaTeX.org`. In addition, specify some keywords, or tags, that can be used to filter the site's content.

Stack Exchange is a commercial network of Q&A sites. Since 2021, it has been owned by **Prosus**, a technology investor. The TeX and LaTeX site, often called **TeX.SE** for short, was founded in 2010 specifically for users of TeX and LaTeX.

This TeX Q&A site has developed into an easily accessible knowledge base:

- Questions are *tagged*. For each question, choose one or more tags that describe the subject. For example, if your question is about a problem with `\label` and `\ref` for equations, select the `cross-referencing` and `equations` tags. This makes it easy to find answers to specific subjects. Specialized experts watch their favorite tags.
- We can vote on answers. Users vote helpful, meaningful answers up while voting down misinformation. This way, the best answer floats to the top. So, you don't have to read through an entire multi-page thread to find the best solution, as is necessary for classical web forums.

Both tagging and voting enhance information access. We can use them to sort and refine search results.

There's another concept called *reputation*. You don't need to worry about that, but you may want to know what it means. Users who post good questions and valuable answers earn reputation points based on the votes they receive.

A certain number of reputation points allows you to go beyond simply asking and answering:

- You can create new tags or retag questions
- Advertising is reduced
- You can edit other users' posts
- You can get access to specific moderation tools

Reputation is a rough measurement of the user's status in the community. A high reputation means greater trust and access to moderation. In that way, the site is community-moderated and shows aspects of a collaborative wiki.

For new users, there is a starter guide: `https://tex.meta.stackexchange.com/questions/1436/welcome-to-tex-sx`.

You can also find further information in the help center at `https://tex.stackexchange.com/help`.

Since TeX.SE is very strict, and questions related to already existing content may quickly be closed without answers, `LaTeX.org` may be a better choice for beginners with any question.

Reddit

Reddit is a collection of discussion forums, each dedicated to a specific topic. These forums are called **subreddits**. There is one for LaTeX at `https://www.reddit.com/r/LaTeX`, where you can ask practical questions, browse earlier discussions, and often get quick responses from other users, known as **redditors**.

As with most forums, posts should be clear, be focused on LaTeX, and include enough information to be useful. Content is voted up or down by the community, so helpful questions and answers usually move to the top. Posts also carry small labels, called **flairs**, that indicate the type of content and can be used to filter posts, making it easier to find what you are looking for.

Facebook

The public Facebook LaTeX user group at `https://www.facebook.com/groups/170883319628439` has more than 9,000 members. It's a place for general LaTeX discussions and comes with the usual Facebook features, including the infamous *Like* button.

Forums in other languages

The following sites are very similar to TeX.SE:

- `https://texnique.fr`: A French Q&A site on LaTeX
- `https://texwelt.de`: A German Q&A site on LaTeX

And, similar to LaTeX.org, `https://golatex.de` is a German LaTeX web forum.

Usenet groups

Around 1980, a long time before the World Wide Web was born, **Usenet** emerged. This is a discussion network that consists of many thousands of groups, so-called **newsgroups**, each dedicated to a specific subject. Unsurprisingly, there are TeX newsgroups.

The most famous one is `comp.text.tex`. The easiest way to access it is to visit `https://groups.google.com/g/comp.text.tex`. Just browse it using its web interface.

Alternatively, you could install a Usenet reader program and connect to a Usenet web server. At this point, you should familiarize yourself with Usenet. A great starting point is its Wikipedia entry at `https://en.wikipedia.org/wiki/Usenet`. There, you will find an introduction, links to necessary software, and further reading.

`comp.text.tex` is the classic TeX discussion board. Then and now, there are distinguished experts reading and posting messages. You can search and browse an archive that reaches back more than 20 years.

There are newsgroups in other languages as well. You could check out the German or French Usenet TeX groups if you understand those languages:

- `https://groups.google.com/g/de.comp.text.tex`
- `https://groups.google.com/g/fr.comp.text.tex`

However, over time, Usenet newsgroups became pretty quiet. Nowadays, users tend to visit web forums and Q&A sites.

Lists of frequently asked questions

Now you know where to ask for help. However, during the long existence of online LaTeX communities, the probability that another user encountered the same problem as you is very high. There are a bunch of questions that keep popping up. If you post such a question, the community member might point you to a **frequently asked questions (FAQs)** page. This refers to a list of answers to these FAQs. The following websites host famous collections:

- **TeX FAQ** (`https://texfaq.org`): This is an FAQ site maintained by the UK TeX Users Group. It contains several hundred FAQs and well-thought-out answers. They are sorted by topic, and that list is still growing and being continuously improved.

- **Visual LaTeX FAQ** (`https://ctan.org/pkg/visualfaq`): This site has a very different approach. The Visual FAQ is a PDF containing hundreds of textual and graphical elements, including tables, figures, lists, footnotes, and math formulas. It is 30 pages long and full of demonstration samples. All document objects, including key positions, are marked and hyperlinked. Just clicking on any marked object leads you to the corresponding TeX FAQ entry. Take a look; it's a fancy interface.
- **MacTeX FAQ** (`https://www.tug.org/mactex/faq.html`): This resource is made for you if you are a Mac user. It covers the installation and use of the MacTeX LaTeX distribution and the popular Mac LaTeX editor TeXShop.
- **AMS-Math FAQ** (`https://www.ams.org/faq`): This lists questions and answers relating to `amsmath`, the most recommended LaTeX math package.
- **LaTeX Picture FAQ** (`https://ctan.org/pkg/l2picfaq`): This was developed to answer many questions about including pictures. It deals, for example, with image file formats, conversion tools, picture manipulation, and the placement of floating figures. The document contains many small code examples and is a handy resource for a LaTeX beginner.

  As it originated in a German LaTeX forum, it's written in German. It has been translated into English, but the translation was not yet published as of 2026.
- **German TeX FAQ** (`https://texfragen.de`): This is a German-language list of answers to frequently asked LaTeX questions, organized by topic.

It's usually a good idea to check an FAQ list before asking in a forum or on a mailing list, which is our next topic.

Mailing lists

Now we return to a traditional medium: electronic mailing lists. They are used for both announcements and discussions. If you subscribe to such a list, you will receive announcements and discussion contributions from other subscribers. You could silently receive and read all the messages, and you could send emails to the list address, which would then be sent to all other subscribers. Before sending a general query to the subscriber list, you should check the FAQs.

Today, many people prefer easily accessible media, such as web forums. However, mailing lists still exist and will be in use as long as email is popular. The following mailing lists may be handy for you:

- **texhax** (`https://tug.org/mailman/listinfo/texhax`): This is a list for general TeX discussion, a companion to `comp.text.tex`, established in the 1980s. It has hundreds of subscribers, many of whom are experts.
- **tex-live** (`https://tug.org/mailman/listinfo/tex-live`): This deals with the TeX Live collection. If you have installed this software distribution, you might be interested in subscribing to get the latest news and read and write about its issues.
- **texworks** (`https://tug.org/mailman/listinfo/texworks`): This supports users of the LaTeX editor TeXworks, which we used in *Chapter 1, Getting Started with LaTeX*. You may subscribe if you decide to use that editor later on and are interested in the latest builds, tricks, scripts, and news.

There are many more mailing lists available at `https://tug.org/mailman/listinfo`. There are more than 60 mailing lists for TeX and LaTeX topics, including bibliographies, hyphenation, PostScript, pdfTeX, and development.

TeX user groups and developers of LaTeX editors and other software often provide mailing lists, especially for announcements. You can read about that on their home pages, but we shall look at some of their websites next.

TeX user group websites

TeX user groups are organizations for people interested in TeX and LaTeX. They provide support for their members but also for TeX and LaTeX users in general. Let's look at a few.

The TeX Users Group

The **TeX Users Group** (**TUG**) is a not-for-profit organization with a very long history. Their website is at `https://tug.org`. Founded in 1980, the TUG always had a significant influence on the development and popularity of TeX. The TUG home page is a portal to the TeX world with links to support, documentation, and software. It hosts an extensive collection of TeX-related internet resources at `https://tug.org/interest.html`. An index and a substantial number of links guide you to helpful material on the internet.

It publishes a journal that appears three times a year, and it holds yearly international conferences. It also hosts the LaTeX Font Catalogue at `https://tug.org/FontCatalogue`, which lists nearly all fonts available for use with LaTeX. About a dozen categories, such as sans-serif, typewriter, and calligraphy fonts, help you find the right font. The fonts are displayed briefly in overviews and extensively, with several styles and math examples. The cherries on the cake are specific code examples.

DANTE

Deutschsprachige Anwendervereinigung TeX e. V. (DANTE) is a large German-language TeX users group that funds projects and provides services for the whole TeX world. The home page, `https://www.dante.de`, is a good starting point for German language users.

DANTE provides financial support for running servers that host LaTeX software, web forums, Q&A sites, FAQs, tools, and more, as mentioned in the introduction to this chapter.

The LaTeX project

The LaTeX3 project team maintains the LaTeX 2e standard and develops the next version of LaTeX. The website, `https://www.latex-project.org`, informs users about their work and LaTeX in general, and publishes news regularly.

Other local user groups

There are many local TeX user groups from various countries in the world listed here:

- `https://tug.org/usergroups.html`
- `https://ntg.nl/lug`
- `https://dante.de/dante-e-v/stammtische`

Their websites often contain material in national languages and further information on the TeX world.

In the following section, we will see where we can get software, tools, and packages.

Websites for LaTeX software and editors

Like most software manufacturers and distributors, free and open source software projects offer information on their home pages.

LaTeX distributions

Today, there are two significant LaTeX distributions, both very modern and comprehensive, plus some descendants:

- **TeX Live** (`https://tug.org/texlive`): This is a cross-platform LaTeX software collection that runs on Windows, macOS, Linux, and other Unix systems
- **MiKTeX** (`https://miktex.org`): This is a very user-friendly and widely used LaTeX distribution for Windows that now also supports Linux and other Unix distributions

- **MacTeX** (https://tug.org/mactex): This is derived from TeX Live and has been explicitly customized for macOS

Most Linux versions provide a customized version of TeX Live in their repositories.

LaTeX editors

There are many LaTeX editors available, from easy-to-use to complex and professional. Most of them offer syntax highlighting, support for various (La)TeX compilers, and other tools such as **BibTeX**, **biber**, **makeindex**, and **PDF previewers**. A list to explore follows.

Cross-platform

These editors support many systems, including Windows, macOS, Linux, and Unix:

- **TeXworks** (https://tug.org/texworks): This is lightweight and comfortable
- **Texmaker** (https://xm1math.net/texmaker): This offers many features
- **TeXstudio** (https://texstudio.org): This is derived from Texmaker and now provides many additional capabilities of its own
- **Emacs** (https://gnu.org/software/emacs): This is extensible and very customizable, though not easy to use for everyone; however, it's great together with **AUCTeX**, which can be found at https://gnu.org/software/auctex
- **Vim** (https://www.vim.org): This is based on commands entered via a text interface, enhanced by the Vim LaTeX-Suite (http://vim-latex.sourceforge.net)
- **VS Code** (**Visual Studio Code**) (https://code.visualstudio.com): It offers excellent LaTeX support with the **LaTeX Workshop** extension (https://marketplace.visualstudio.com/items?itemName=James-Yu.latex-workshop), including compilation, preview, and forward/inverse search

Of course, the Overleaf online compiler and editor are cross-platform as well.

Windows

In addition to the cross-platform editors, there's the powerful, popular shareware LaTeX editor **WinEdt**. You can download it at https://www.winedt.com. DANTE offers members a discounted license. Also, see the WinEdt community site at http://www.winedt.org.

Linux

Besides all the cross-platform editors, there are the following:

- **Kile** (`https://kile.sourceforge.io`): This is very powerful and designed for the KDE window system, but it also runs on other window managers, such as GNOME, if KDE libraries are installed
- **gedit** (`https://wiki.gnome.org/Apps/Gedit`): This is the lightweight GNOME standard editor, and there's a LaTeX plugin: `https://wiki.gnome.org/Apps/Gedit/LaTeXPlugin`
- **GNOME-LaTeX** (`https://gitlab.gnome.org/World/gedit/enter-tex`): This is another GNOME-based LaTeX editor, formerly called **LaTeXila**

On Linux, we usually choose an editor that fits the selected window manager, KDE or GNOME, or a universal cross-platform editor.

macOS

TeXShop is a very popular Mac LaTeX editor: `https://pages.uoregon.edu/koch/texshop/`.

This editor is believed to have led many new users to LaTeX because of its outstanding usability. The TeXworks editor is modeled on TeXShop.

The visual editor LyX

`https://www.lyx.org` is the home page of the cross-platform editor **LyX**, which looks and feels like a word-processing software but is built on LaTeX. It combines an easy-to-use graphical user interface with the power and structure of LaTeX. You can develop documents mainly using LyX's toolbars and menus, but you may insert LaTeX code at any point.

The LyX wiki offers extensive documentation at `https://wiki.lyx.org`.

On the LyX home page, you will find links for download, news, and support. As LyX is very popular, there is a LyX dedicated support forum at `https://latex.org`.

CTAN – the Comprehensive TeX Archive Network

The network at `https://ctan.org` consists of many servers worldwide that host the most extensive collection of TeX-related material. CTAN serves as a repository for installing and updating LaTeX distributions such as TeX Live.

On the CTAN home page, you will find search features, or you may start browsing the archive directories. We can see nearly every serious LaTeX package in this archive.

In the following section, we will look at websites that showcase examples, images, and code.

Graphics galleries

There are showcase sites on the internet, especially for creating graphics with TeX:

- `https://texample.net` is a TikZ example gallery with hundreds of examples and complete source code, browsable by topic
- `https://tikz.net` is another TikZ picture gallery with source code
- `https://tikz.org` is the website of the book *LaTeX Graphics with TikZ*, with a lot of examples
- `https://pgfplots.net` focuses on plots in 2D and 3D using the `pgfplots` package
- `https://asymp.net` shows how to plot in 2D and 3D using the **Asymptote** language within LaTeX
- `https://feynm.net` demonstrates how to generate **Feynman diagrams** in LaTeX
- `https://latex-cookbook.net` is the website for the *LaTeX Cookbook*, with a gallery of code examples and output
- `https://latexguide.org` is the website for this book, *LaTeX Beginner's Guide*, with a gallery of examples and further information

These websites allow visual browsing of LaTeX graphics documents, sorted by topic, with complete source code and explanations.

Now, let's get personal; we'll turn to user blogs.

LaTeX blogs

Are you interested in LaTeX news and expert opinions? Then, LaTeX blogs may supply you with current LaTeX information:

- `https://texblog.net` is my personal blog, where I write LaTeX-related news, offer tips and tricks, and provide a structured link collection sorted by subject
- `https://www.texdev.net` is written by Joseph Wright, a member of the LaTeX project and author of various LaTeX tools
- `https://tex-talk.net` is a LaTeX blog with an interesting section full of interviews with LaTeX power users and developers
- `https://latex.net` is primarily an article database with accumulated know-how over many years, but it also publishes news posts like a blog

- `https://tex.social` is a blog aggregator that collects RSS feeds from over 30 LaTeX-related blogs and social networks

But even faster news is on Twitter, so let's look at that next.

X, formerly Twitter

Recommended X accounts to subscribe to are as follows:

- `@TeXUsersGroup` (`https://x.com/TeXUsersGroup`): This is the TUG account with news and CTAN updates
- `@overleaf` (`https://x.com/overleaf`): This is the account for the Overleaf company, which offers an online compiler and editor
- `@tex_tips` (`https://x.com/tex_tips`): This account sends daily LaTeX tips to its subscribers
- `@TeXgallery` (`https://x.com/TeXgallery`): This is my account that also represents the LaTeX.org forum

Follow the hashtag `#TeXLaTeX` to get the latest news about TeX and LaTeX: `https://x.com/search?q=%23TeXLaTeX`.

Summary

While you have learned the fundamentals of LaTeX in this book, this chapter provides an overview of additional online resources.

Now you know about finding and downloading LaTeX software, accessing the worldwide LaTeX community knowledge, getting the latest news from blogs, and asking questions online if you encounter a problem that you cannot solve by yourself.

TeX friends will welcome you on any community website. As you have learned much in this book, you may soon become an experienced LaTeX user who supports LaTeX novices.

The rise of **artificial intelligence** (**AI**) has changed how people get help. Today, many users simply ask an AI chatbot such as ChatGPT, Claude, Gemini, Grok, or Microsoft Copilot instead of posting a question on a forum. We'll take a closer look at this in the next chapter, together with a few more advanced topics.

16

Exploring Technology: From Engines to AI

After working through 15 chapters and building a solid understanding of LaTeX, it's a good moment to step back and look at what's under the hood. In this final chapter, we'll explore the technology behind LaTeX itself, take a look at TeX specifically and at modern developments, and point to further books and resources worth reviewing.

This chapter covers the following topics:

- Looking at TeX engines
- Understanding TeX formats
- Selecting engine and format
- Using artificial intelligence
- Further reading

As they say before a race at NASCAR or IndyCar: drivers, start your engines.

Looking at TeX engines

A TeX **engine** is the program that performs the actual typesetting. It reads your input file, understands the commands, expands macros, breaks text into lines and pages, and produces the final output. When you press **Typeset** or **Compile**, the engine turns your source code into a finished document.

Different TeX engines exist because the requirements for typesetting changed over time. The original TeX was designed long before **Unicode** became the standard for representing text, with support for many writing systems. At the same time, modern fonts usually come in the **OpenType** format. OpenType fonts support large character sets and advanced typographic features, such as real small caps, ligatures, stylistic variants, and language-specific glyphs. Newer TeX engines were designed with these capabilities in mind so that they can handle Unicode text and OpenType fonts much more naturally.

Let's go through the available engines in a bit more detail.

The classic TeX engine and DVI output

The original TeX engine did not create PDF files. Instead, it produced a `.dvi` file. **DVI** is short for **Device Independent**. This file describes the document layout without targeting any specific printer or screen. Separate programs can convert the `.dvi` file into printable or viewable formats.

Internally, TeX builds pages from small building blocks, called **boxes**, and uses flexible spacing, often called **glue**, to achieve high-quality line and page breaks. In practice, DVI files were rarely the final result. Users usually converted them into PDF using additional tools. A common workflow was to first convert the DVI file to a **PostScript** file using the **dvips** tool. PostScript is a page description language that printers and previewers understand well. This PostScript file was then converted into a PDF using the **ps2pdf** program.

The ps2pdf tool is part of **Ghostscript**, a widely used software suite that processes and converts PostScript and PDF files. In some setups, users can also use **dvipdf** or the more modern **dvipdfmx** tool, which converts a DVI file directly to PDF without an intermediate PostScript step.

Most users never call dvipdf or dvipdfmx directly. Editors and TeX distributions automatically choose the right tool. You usually just press **Typeset**, and the correct conversion step happens in the background.

The classic TeX engine supports external images only in the EPS format, rather than common formats such as PNG or JPEG, as you saw in *Chapter 6, Including Images.*

If you want to see how TeX started and what plain TeX looks like at its core, have a look at Donald Knuth's *The TeXbook* at `https://ctan.org/pkg/texbook`. You can compile it yourself or buy the book. Another great book is *TeX by Topic* by Victor Eijkhout, freely available at `https://www.eijkhout.net/tex/tex-by-topic.html`.

eTeX extends the classic TeX engine with a small set of features that primarily help with writing more robust macros and improve some internal behavior. For most users, eTeX feels just like classic TeX, so existing documents continue to work unchanged. Today, eTeX's extensions are part of all modern TeX engines, which we'll look at next.

pdfTeX: today's workhorse

pdfTeX is the engine many LaTeX users start with. It produces PDF files directly and supports common image formats such as PNG and JPEG.

It comes with excellent support for classic typography, including microtypographic features. Its main limitation is font handling. pdfTeX is limited to traditional PostScript fonts and rarely used bitmap fonts; it does not work directly with modern system fonts. Even so, it is still widely used because it is reliable, fast, and more than sufficient for most documents.

For details about pdfTeX, its features, and its limitations, start here:

- Official pdfTeX project page: `https://tug.org/applications/pdftex`
- The pdfTeX manual on CTAN: `https://ctan.org/pkg/pdftex`

XeTeX: Unicode and system fonts

XeTeX is designed to work with Unicode text and modern fonts. Internally, it uses Unicode everywhere and lets you access system fonts directly, which is a major improvement over pdfTeX.

This makes XeTeX a good choice for multilingual documents or texts that use non-Latin scripts. It is easy to use and solves font issues cleanly, but it does not focus on programmability or advanced automation.

The name was chosen to suggest an *eXtended* version of eTeX. Since it supports right-to-left scripts such as Arabic and Hebrew, a name that reads the same in both directions felt like a fitting and playful choice.

If you want to learn more about XeTeX, Unicode support, and system fonts, visit the XeTeX project page at `https://xetex.sourceforge.net`.

LuaTeX: extensibility and programmability

LuaTeX takes things a step further by embedding the **Lua** programming language directly into the engine. This allows Lua code to interact with the typesetting process and to be used by authors.

That becomes useful when documents are no longer static. With LuaTeX, you can generate content dynamically, process data, or implement complex logic during compilation. Like XeTeX, it fully supports Unicode and modern fonts. While it compiles a bit slower than the other engines, it's better maintained than XeTeX and more closely integrated, and its features and programmability make it the most flexible TeX engine available today.

If you use LuaLaTeX and notice that **LuaHBTeX** is being used, that's simply a LuaTeX variant: the letters HB stand for **HarfBuzz**, a text-shaping engine that turns Unicode text into correctly positioned glyphs and handles complex scripts, ligatures, and modern font features.

Let's take a brief look at how we can use Lua in our documents. We'll go through a very small example that implements the **Euclidean algorithm** to calculate the **greatest common divisor (GCD)** of two integer numbers:

1. Start a new document. Use any document class you like:

   ```
   \documentclass{article}
   ```

2. Load the `luacode` package:

   ```
   \usepackage{luacode}
   ```

3. Use a `luacode` environment to implement a Lua function GCD with two arguments, a and `b`:

   ```
   \begin{luacode}
     function GCD(a,b)
       if b ~= 0 then
           return GCD(b, a % b)
       else
           return a
       end
     end
   \end{luacode}
   ```

4. Start the document:

   ```
   \begin{document}
   ```

5. Write some text that uses the `\directlua` command to do a calculation:

```
The greatest common divisor of 96 and 36 is
\directlua{tex.sprint(GCD(96,36))}.
```

6. End the document:

```
\end{document}
```

7. Compile the document. You will receive this output:

The greatest common divisor of 96 and 36 is 12.

The `luacode` package provides a safe way to embed Lua code directly in a LaTeX document without worrying about how TeX would handle characters and spacing. The environment with the same name executes Lua code, and you can access it using the `\directlua` command.

Regarding that function: as long as `b` is not `0`, the function runs again with the `b` and `a%b` arguments, which is what remains when `a` is divided by `b`. This keeps the same GCD and reduces the numbers each time until the process finally stops. A good explanation can be found at `https://en.wikipedia.org/wiki/Greatest_common_divisor`.

`tex.sprint` sends text from Lua back to LaTeX, as if you had typed that text directly into your document.

It's really a simple document to demonstrate that you can do even complex calculations using Lua right within LaTeX. Here you can see examples I created using more complex calculations for plotting fractals, chaotic systems defined by differential equations, and other algorithms: `https://pgfplots.net/tag/luatex`.

For more information, visit `https://www.luatex.org` and `https://www.lua.org`.

LuaMetaTeX: combining TeX, Lua, and MetaPost

LuaMetaTeX is a TeX engine that tightly integrates Lua, Unicode, modern font support, and the **MetaPost** graphics language functionality. As the successor to LuaTeX, it is the engine used by the modern **ConTeXt** format, which we will learn about in the next section.

There are also engines for specific languages. The **pTeX** engine is for Japanese typesetting, and **upTeX** is its modern Unicode-based successor.

Understanding TeX formats

A TeX **format** is a predefined set of macros built on top of a TeX engine. While the engine handles low-level typesetting, the format defines the commands that structure documents and that you, as an author, use.

You can think of it like this: the engine is the machinery, while the format is the user interface. Let's look at the most important formats you'll encounter.

Plain TeX

Plain TeX is the original format; it's small and very close to the underlying TeX engine. It gives you full control but very little convenience. There is no built-in document structure, so even simple things such as sections and references require more work. Today, Plain TeX is mainly of historical interest and used by a few experts who want maximum control.

LaTeX

LaTeX builds on Plain TeX and adds a rich set of high-level commands for document structure. You tell it what something is, such as a section, a figure, or a table, and LaTeX takes care of the layout. You've seen it in this book; no more words needed.

LaTeX has a huge ecosystem of packages and works well for both simple and complex documents. Combined with the name of the underlying engine, we call the different flavors **pdfLaTeX**, **XeLaTeX**, **LuaLaTeX**, **pLaTeX,** and **upLaTeX**.

ConTeXt

ConTeXt is a TeX format with a different approach from LaTeX. It's a consistent, all-in-one system where many features are built in rather than added through separate packages.

It offers strong layout and typography features right away, but it has a very small user base and ecosystem. Users often choose ConTeXt for fine control over design within a single framework.

One of the most visible differences between ConTeXt and LaTeX is how environments are written. In LaTeX, environments use `\begin{...}` and `\end{...}`. In ConTeXt, the same idea is expressed with matching `\start...` and `\stop...` commands.

Let's have a quick look at a sample ConTeXt document similar to what we wrote in *Chapter 12*:

```
\setuppapersize[A4]
\starttext
\startchapter[title={Equations}, reference=ch:eq]
\startsection[title={Quadratic equations}, reference=sec:quad]
A quadratic equation is an equation of the form
\placeformula[eq:quad]
\startformula
  ax^2 + bx + c = 0
\stopformula
where $a$, $b$, and $c$ are constants and $a\neq 0$.

As shown in \in{equation}[eq:quad], quadratic equations are
discussed in \in{section}[sec:quad] of \in{chapter}[ch:eq].
\stopsection
\stopchapter
\stoptext
```

While LaTeX uses commands for chapters and a generic environment system, ConTeXt uses the same start-and-stop pattern everywhere, including high-level structures such as chapters and sections. Just like in LaTeX, these blocks can be nested.

If you want to explore ConTeXt further, the best starting points are these:

- The official ConTeXt website at `https://www.contextgarden.net`
- The ConTeXt manuals at `https://www.pragma-ade.nl/general/manuals`

Selecting engine and format

First, as a practical guideline:

- Use **pdfLaTeX** for classic documents with standard fonts and a stable workflow
- Choose **LuaLaTeX** if you want system fonts or need Lua programming
- Use **XeLaTeX** if a template requires it or if its right-to-left features fit your needs better; development today focuses more on LuaLaTeX
- Try **ConTeXt** when you want Lua-based typesetting with integrated layout control and are ready to step beyond LaTeX with a different syntax, that's still TeX

LaTeX editors let you easily select the engine. For example:

- In **TeXworks**, choose the engine from the dropdown next to the **Typeset** button, as you see in *Figure 10.19*
- In **TeXstudio**, go to **Options** | **Configure TeXstudio** | **Build** | **Default Compiler**
- In **Overleaf**, open **Menu/File** | **Settings** | **Compiler** and select **pdfLaTeX**, **XeLaTeX**, or **LuaLaTeX**

If you work on the command line, in a terminal window, select the engine by running its (lowercase) name, like one of these commands:

```
pdflatex filename.tex
xelatex filename.tex
lualatex filename.tex
context filename.tex
```

Using artificial intelligence

Artificial intelligence (**AI**) is software that does not follow fixed instructions like traditional programs. Instead, it learns by analyzing large amounts of data. It processes information, including your input, using logic, statistics, and trained algorithms.

AI has made rapid progress in recent years and has become widely known with the rise of **ChatGPT**. ChatGPT is a **chatbot**: you enter a question or instruction, and it responds based on what it has learned during training. It became popular because interacting with it feels very natural, almost like talking to a person. You can ask follow-up questions, refine answers, and even adjust things such as tone, writing style, or level of detail.

Sometimes the output is exactly what you need. Sometimes it only sounds right, and errors can happen. But quite often, the result is good enough to be genuinely useful. So, let's see how we can make it work for us.

There are many AI chatbots available today, created by different companies and research groups. Most of them work similarly; the main differences are the interface and the underlying training data.

Here's a list of currently popular AI chatbots and where you can find them:

- **ChatGPT** by OpenAI: `https://chat.openai.com`
- **Claude** by Anthropic: `https://claude.ai`

- **Gemini** by Google: `https://gemini.google.com`
- **Copilot** by Microsoft: `https://copilot.microsoft.com`
- **Perplexity** by Perplexity AI: `https://www.perplexity.ai`
- **Meta AI** by Meta: `https://www.meta.ai`
- **Grok** by X.ai: `https://x.ai`

All of these are useful, and there is no single "best" choice. Personally, I recommend using ChatGPT. It is easy to get started with and widely used, and it works very well for text-focused tasks such as generating LaTeX source code.

ChatGPT offers a free version that's perfectly fine to begin with. It comes with usage limits and gives you access to smaller models with fewer features. There's a paid version that provides access to more capable models, faster responses, higher usage limits, and better reasoning. If you start using it frequently, it can be worth it.

In the *LaTeX Coobook, Chapter 13, Using Artificial Intelligence with LaTeX*, I wrote various step-by-step instructions along with ChatGPT responses. Here, let's keep it a bit shorter and focus on general advice, since the technical interfaces and features of AI bots change rapidly.

Here's how to use ChatGPT:

1. Open the website, `https://chat.openai.com`, in your browser.
2. Click **Sign up** to register with an email address and password, or quickly use a Google, Microsoft, or Apple ID account.
3. OpenAI will send you an email to confirm your address. Click the **Verify email address** button in that message.
4. After creating your account, use **Log in** to continue.
5. Once you're logged in, you're ready to start chatting. You can ask questions or give instructions, and ChatGPT will respond.

First, let's talk about your input to the AI.

Working with AI efficiently

When you use an AI chatbot, the text you type into it is called a **prompt**. That's the input you give to an AI system, which can be a question, an instruction, or a piece of text to analyze. It triggers the AI to generate a response.

Prompts are the key to efficient use. You can be precise: enter topic, context, purpose, situation, audience level, and any constraints that matter. The clearer the prompt, the better the result. It tells the AI what to do, for whom, and under which conditions.

Here are examples of strong prompts:

- **Be explicit**: "*Explain how LaTeX float placement works, focusing on precise positioning, and use clear and easy language for advanced beginners.*"
- **Provide context**: "*Explain how biblatex and biber work together, for a user with basic LaTeX knowledge, working on macOS, including restrictions and compatibility aspects.*"
- **Define constraints**: "*Summarize the following text in under 150 words, continuous prose, active voice, friendly tutorial tone without enthusiasm, avoid bullet points, prefer commas or parentheses over em dashes.*" (then paste the text)
- **Iterate**: "*Rewrite your previous answer, make it shorter and slightly more formal, while keeping all technical details.*"

You can certainly ask in short, spontaneous phrases, but longer and more deliberate prompts usually lead to better results, especially if you plan to reuse them again and again.

Let's look at some ideas of what you can enter.

Asking LaTeX-related questions

You can ask any question about LaTeX that comes to mind, such as the following:

- "*Which document class is suitable for a presentation? Give reasons, and show me a complete small LaTeX code example.*"
- "*What is the difference between LuaLaTeX and pdfLaTeX? Provide decision guidance.*"

You can see that it's recommendable to add what you expect from an answer. The results are often surprisingly good, though the AI may struggle with niche topics. Once you get an answer, you can ask follow-up questions for clarification, alternatives, or a deeper explanation.

Suggesting packages and alternative solutions

You can ask the AI assistant to recommend suitable LaTeX packages and outline alternative approaches to solve a problem. When several solutions exist, this helps you compare options and decide which one best fits your document, often faster than searching through manuals or forums. You can start with a question like this: "*Which package is best for handling references and citations in a long document? Compare the main options and explain when to use each.* "

Generating LaTeX code

Give a chatbot a clear instruction, and it can generate LaTeX code for you, saving time and effort. This is especially useful for complex tables, where writing tabular code by hand can be tedious. You can describe what the table should look like, and the chatbot gives you a solid starting point. It can also help create BibTeX entries from basic information, set up figure environments, or draft longer code blocks that you can easily adjust. This works for almost any complex syntax structure, including presentations with overlays: you describe what you need, get complete code, and then refine it. AI-generated code may still require adjustments before it compiles cleanly or fits your document perfectly.

Here's a strong example prompt:

"Generate a minimal but complete LaTeX document that creates a professional-looking table suitable for a scientific thesis. The table should have five columns, alternating row colors, aligned numbers in the first column, a caption, and a reference label. Use well-established packages and briefly explain your design choices."

The prompt clearly defines the layout, sets the context, asks for package choices, and requests a short explanation so you understand what the code actually does.

Explaining error and warning messages

LaTeX error and warning messages can look cryptic at first glance. An AI assistant can translate them into plain language. You can paste the message into the chat and ask what it typically means, where it usually occurs, and what part of your document you should inspect.

It can also list common causes and suggest practical steps to resolve the issue. This makes debugging faster and far less frustrating.

Here's an example prompt:

"I get the following LaTeX error. Explain what this error means in simple terms, what usually causes it, and give step-by-step guidance on how to locate and correct it. Error message: Undefined control sequence. "

I admit that I often keep my prompts much shorter. I'm usually lucky with the results, but a more detailed prompt clearly increases the chances of getting exactly what I need.

Troubleshooting and debugging

An AI assistant can walk you through the troubleshooting step by step. Describe what goes wrong, what you expected instead, and paste a piece of code. Based on that, it can suggest likely causes and practical corrections, helping you narrow the problem down quickly.

Think of it as a discussion partner for debugging. You paste the relevant code, include the exact error or warning message, and briefly explain the context. From there, you can iterate: test the suggestion, report back, and refine the solution.

Improving and polishing existing text

An AI assistant can help you refine text you've already written. Paste a paragraph and ask for clearer wording, smoother flow, or simpler explanations. It can spot grammar issues, reduce repetition, and adjust tone while preserving your original meaning.

This is especially helpful when you want consistency across a longer document. You can specify the desired tone in advance, such as informal, technically precise, beginner-friendly, or neutral, and then ask for a revision based on that style.

Here's an example prompt:

"*Revise the following paragraph. Keep the technical meaning unchanged. Use a clear, informal tutorial tone. Avoid enthusiastic and exaggerated language. Prefer active voice and short sentences. Reduce repetition and improve logical flow.*" (Paste your paragraph here)

Note that this helps control the tone, since AI responses tend to sound overly enthusiastic by default.

You can also use an AI assistant for translation. I've often used ChatGPT to communicate with Japanese LaTeX colleagues, translating messages and then asking it to explain the wording step by step so I could verify that everything was accurate.

For grammar checking, I also frequently rely on ChatGPT. However, for pure grammar and style corrections, I find specialized tools such as **Grammarly Pro** more efficient. Sometimes I use both and then decide whether to keep my original phrasing or adopt one of the suggested revisions.

Brainstorming structure, titles, and section ideas

An AI assistant can be very helpful when you're stuck or want to plan faster. If the page is still blank, you can ask it to suggest a document structure, section outline, or possible titles. This helps you overcome writer's block and get an initial framework. You don't have to accept the result as is, but it often provides a solid starting point for moving forward.

Here's an example prompt:

"*I'm writing a document about AI support in LaTeX-based technical writing. Propose a detailed outline for one chapter on AI-assisted debugging of LaTeX documents. Include suggested section titles, a logical flow, key arguments, and possible examples or experiments. Keep it academically rigorous but practical.*"

Using this, ChatGPT 5.2 gave me an outline of 12 sections with key points and explanations that would give me a good start to write my own thoughts.

Using AI responsibly and carefully

I don't encourage using AI to generate content outright. The responsibility remains with you as the author. Suggestions, rephrasing, or structural ideas should be treated as drafts, not as finished results. Review everything carefully, verify the facts, ensure completeness, and make sure the wording reflects your intent and fits your audience. AI can speed up parts of the process and help you think, but you decide what ultimately goes into the document.

Be cautious when sharing sensitive or confidential information with AI tools. Text you submit may be stored or processed by the provider and become accessible to others. Avoid uploading private data, credentials, or unpublished material. If you use AI support in such cases, anonymize names and details, and reduce the content to what is strictly necessary.

Looking at the future of AI and LaTeX

It's hard to predict the future in detail, but a few trends are already visible.

First, AI chatbots are changing how people seek help. For many everyday questions, they now compete directly with traditional web forums and Q&A sites. Bots respond instantly, remain patient, and are usually technically correct. That makes them more attractive than asking humans and waiting for replies. Community platforms still play an important role, but some commercial Q&A sites are under pressure as traffic and revenue decline. I'll continue running community-driven forums like LaTeX.org for human discussion, yet the landscape is clearly shifting.

Second, AI is becoming part of the tools we already use. Editors and platforms are increasingly integrating AI directly into the writing workflow. Overleaf is one example, and new tools are appearing.

One such tool is Prism (`https://prism.openai.com`), a web-based LaTeX editor with built-in AI assistance. It runs entirely online and lets you write, edit, and collaborate without a local LaTeX setup. The key difference is that the AI works inside the editor and understands your document structure, source code, and context. You can ask for wording improvements, structural suggestions, or explanations without leaving LaTeX.

For researchers and technical writers, this can save time on setup, formatting, and revision. AI can help polish text and speed up the writing process while you stay in control of the document.

Again, an important reminder: when you share text with an AI service, that content may be stored, reused for training, or appear in responses to other users. If that's acceptable for your use case, fine. If not, be cautious.

Further reading

We have come to the end of this book. In each chapter, I wanted to tell you even more. I often referred to other books I wrote, so let's look at why and what you can find in them.

LaTeX Cookbook: over 100 practical and advanced LaTeX solutions

After I wrote the first edition of *LaTeX Beginner's Guide*, it was clear to me that there's much more that a beginner's tutorial could cover. So, I worked on a set of examples for various document types, including CVs, letters, leaflets, and posters, all beyond the scope of this book, and I wrote a lot of advanced advice on various topics, with specific solutions. That's when I created the *LaTeX Cookbook*. It differs from *LaTex Beginner's Guide*, which introduces concepts step by step and explains fundamentals in a learning sequence. A cookbook assumes basic familiarity and focuses on concrete solutions to specific problems. Each section is largely self-contained, with short explanations followed by practical examples that can be adapted directly to real documents.

Let me summarize this book. It starts with document classes, text tuning, and font selection, showing how different document types are configured and how typographic decisions affect the output. The next chapters focus on core content elements: tables, images, graphics, and design components. They cover practical techniques for layout, alignment, floating objects, and visual enhancement beyond standard text.

Document-wide structure and output follow, including tables of contents, lists, bibliographies, glossaries, indexes, and PDF features such as metadata, forms, merging, and optimization. Advanced sections address mathematical typesetting and scientific applications, including formulas, theorems, plots, diagrams, and examples from science and technology.

The final chapters cover external support and modern workflows, with guidance on online resources and the controlled use of AI tools to assist LaTeX work.

You can find detailed information at `https://latex-cookbook.net/contents`. Due to high demand, the book was translated and published in Japanese; the book's website is `https://tex.jp`.

LaTeX Graphics with TikZ: drawing images

I wrote the *LaTeX Cookbook* with many graphics examples, including various diagram types, which are easy to create. I understood that, for a serious coverage of LaTeX's features for drawing images, a whole new book is needed. So, I wrote the first-ever book on drawing pictures in LaTeX with **TikZ**.

This book introduces TikZ step by step, from basic concepts to advanced graphics and automation. It starts by explaining what TikZ is, how it fits into LaTeX, and how drawings are built from coordinates, shapes, and color. Early chapters establish the core syntax and drawing model. The focus then moves to structured graphics based on nodes and edges. You learn how to place and align nodes, connect them with lines and arrows, and control labeling, styles, and direction.

Next, the book introduces abstraction and reuse. Styles, scopes, and reusable picture components are combined with higher-level structures, such as trees, graphs, mind maps, and matrix layouts. Visual refinement follows, covering filling, clipping, shading, path decorations, layers, overlays, and transparency, including the integration of TikZ graphics with the rest of the document.

Advanced chapters treat TikZ as a calculation and geometry engine, showing coordinate calculations, intersections, loops, transformations, and techniques for drawing smooth curves. The final sections focus on data visualization and diagram generation, including 2D and 3D plots and package-based approaches for building complete diagrams, before concluding with creative examples from the TikZ community.

You can find more information about the book at `https://tikz.org`. The book has also been published in Japanese; its website is `https://tikz.jp`.

The LaTeX Companion

The LaTeX Companion, in its third edition, is published in two volumes and serves as an in-depth reference for LaTeX and its ecosystem.

Volume I focuses on core document construction. It covers document structure, text formatting, page layout, tables, floats, graphics, and font selection, explaining how standard LaTeX features and widely used packages work and interact.

Volume II addresses advanced and specialized topics. It includes extensive coverage of mathematical typesetting, symbols, Unicode, and multilingual support, bibliographies and citations, indexes, customization, and selected aspects of LaTeX programming and package design.

This book is not a beginner tutorial and not recipe-driven; it's best suited as a long-term reference for experienced LaTeX users who want detailed explanations and authoritative guidance.

Summary

We stepped back from everyday LaTeX use and looked at what runs underneath our documents. You learned what TeX engines do, why different engines exist, and how TeX formats build on them.

We also looked at how AI can support the LaTeX workflow, helping with code generation, error explanations, troubleshooting, text refinement, and planning, while still letting you stay in control.

With all 16 chapters completed, you now have a solid foundation. We closed by looking at further books if you want to dive deeper.

17

Unlock Your Exclusive Benefits

Your copy of this book includes the following exclusive benefits:

Follow the guide below to unlock them. The process takes only a few minutes and needs to be completed once.

Unlock this book's free benefits in 3 easy steps

Step 1

Keep your purchase invoice ready for *Step 3*. If you have a physical copy, scan it using your phone and save it as a PDF, JPG, or PNG.

For more help on finding your invoice, visit `https://www.packtpub.com/en-us/unlock?step=1`.

Note: If you bought this book directly from Packt, no invoice is required. After *Step 2*, you can access your exclusive content right away.

Step 2

Scan the QR code or go to `packtpub.com/unlock`.

On the page that opens (similar to *Figure 17.1* on desktop), search for this book by name and select the correct edition.

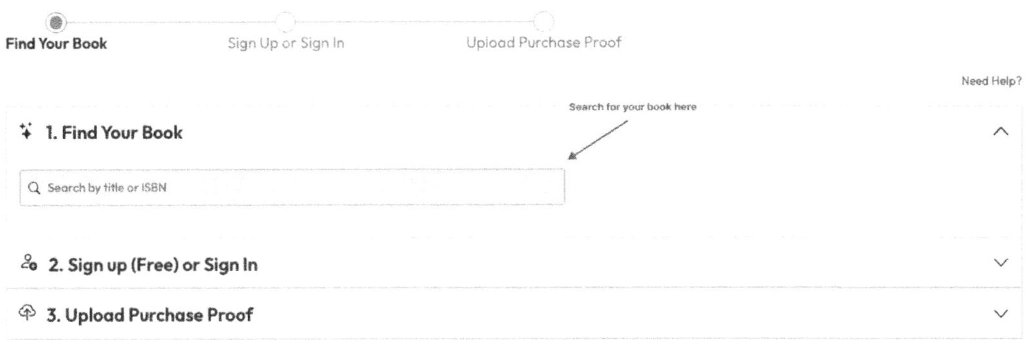

Figure 17.1: Packt unlock landing page on desktop

Step 3

After selecting your book, sign in to your Packt account or create one for free. Then upload your invoice (PDF, PNG, or JPG, up to 10 MB). Follow the on-screen instructions to finish the process.

Need Help

If you get stuck and need help, visit `https://www.packtpub.com/unlock-benefits/help` for a detailed FAQ on how to find your invoices and more. This QR code will take you to the help page.

Note: If you are still facing issues, reach out to `customercare@packt.com`.

Index

C

F

G

M

Q

R

S

T

U

V

W

X

Z

Other Books You May Enjoy

If you enjoyed this book, you may be interested in these other books by Packt:

LaTeX Graphics with TikZ

Stefan Kottwitz

ISBN: 9781804618233

- Understand the TikZ language and how to use its libraries and packages
- Draw geometric shapes with text and add arrows, labels, and decorations
- Apply transformations and use transparency, shading, fading, and filling features
- Define styles and program with loops to streamline your code
- Build trees, graphs, and mind maps and draw easy curves with smooth transitions
- Produce block diagrams and flow charts to visualize process steps
- Generate line charts and bar charts to showcase your data
- Plot data sets and mathematical functions in two and three dimensions

LaTeX Cookbook

Stefan Kottwitz

ISBN: 9781835080320

- Utilize various document classes and incorporate bibliography, glossary, and index sections
- Handle arranging and annotating images with ease
- Create visually appealing tables and learn how to manage fonts efficiently
- Generate diverse and colorful graphics, including diagrams, flow charts, bar charts, trees, and both 2D and 3D plots
- Solve writing and drawing tasks across various scientific disciplines
- Optimize PDF output, enhancing it with metadata, annotations, popups, animations, and fill-in fields
- Leverage ChatGPT to improve content and code

Packt is searching for authors like you

If you're interested in becoming an author for Packt, please visit `authors.packt.com` and apply today. We have worked with thousands of developers and tech professionals, just like you, to help them share their insight with the global tech community. You can make a general application, apply for a specific hot topic that we are recruiting an author for, or submit your own idea.

Share your thoughts

Now you've finished *LaTeX Beginner's Guide, Third Edition*, we'd love to hear your thoughts! Scan the QR code below to go straight to the Amazon review page for this book and share your feedback or leave a review on the site that you purchased it from.

`https://packt.link/r/180580457X`

Your review is important to us and the tech community and will help us make sure we're delivering excellent quality content.

Made in the USA
Las Vegas, NV
27 March 2026